The Urban Book Series is a resource for urban studies and geography research worldwide. It provides a unique and innovative resource for the latest developments in the field, nurturing a comprehensive and encompassing publication venue for urban studies, urban geography, planning and regional development.

The series publishes peer-reviewed volumes related to urbanization, sustainability, urban environments, sustainable urbanism, governance, globalization, urban and sustainable development, spatial and area studies, urban management, transport systems, urban infrastructure, urban dynamics, green cities and urban landscapes. It also invites research which documents urbanization processes and urban dynamics on a national, regional and local level, welcoming case studies, as well as comparative and applied research.

The series will appeal to urbanists, geographers, planners, engineers, architects, policy makers, and to all of those interested in a wide-ranging overview of contemporary urban studies and innovations in the field. It accepts monographs, edited volumes and textbooks.

Indexed by Scopus.

More information about this series at http://www.springer.com/series/14773

Sotir Dhamo

Understanding Emergent Urbanism

The Case of Tirana, Albania

Springer

Sotir Dhamo
Polis University
Tirana, Albania

ISSN 2365-757X ISSN 2365-7588 (electronic)
The Urban Book Series
ISBN 978-3-030-82730-4 ISBN 978-3-030-82731-1 (eBook)
https://doi.org/10.1007/978-3-030-82731-1

This Springer imprint is published by the registered company Springer Nature Switzerland AG
The registered company address is: Gewerbestrasse 11, 6330 Cham, Switzerland

Acknowledgments

I gratefully acknowledge the support and guidance of many people who helped me during the process of writing this book. First of all, I am grateful to my colleagues and friends, Besnik Aliaj and Dritan Shutina, who unceasingly encouraged me to write this book and, most importantly, boosted my confidence in bringing to a conclusion the things that have been floating in my mind for years. Without their intellectual, moral, and professional inspiration, this book could not have been the same.

I would especially like to thank my colleague, Antonino Di Raimo, for consistently provoking me to move away from the conventional ways by which we view, plan, and design our cities. Wandering with him in the informal and formal areas of Tirana (in an attempt to truly understand the city) and his dissatisfaction over the possibility of only analyzing through conventional categories a city, where the energy of people was visible in the air, had a striking effect on me. His dissatisfaction became my frustration, which then motivated me to move forward with this topic.

I am most grateful to Ledian Bregasi and Valerio Perna for offering their wisdom and their serene nature. They have been patient and excellent interlocutors during difficult moments. I also thank them for their great support in the series of CompleX-City experimental lab activities organized in the framework of the research for this book wherein, among others, we outlined the first idea of a potential modeling process. The support provided by Antonino, Ledian, and Velerio was crucial during this part of my work.

Many thanks also to my colleagues, Ermal Hoxha and Llazar Kumaraku, for being such passionate and tireless collaborators in the site planning analyses and urban design classes in which we experimented and tested a great deal of ideas deemed valuable for this research. Many thanks to Endrit Marku, who generously offered his talent for the artwork interpretations of the analyzed patterns, and to Artan Kacani, with whom I unceasingly debated and exchanged information, especially regarding the topic of informality.

I would also like to sincerely thank the group of students who participated in the CompleX-City experimental lab and all other students—past and future—who have been and always will be an immense source of inspiration for continuous learning during my academic career. The book was written with them in mind.

I am most grateful to Co-plan for being a real open laboratory and for giving me access to its historic archives containing unique photographic and written documents of urban and social processes, which characterized Albania after the collapse of the dictatorship. I am also extremely grateful to the National Technic Archive (NTA), which made available the photographic collection as well as the historic maps and written documents covering the period from the beginning of the past century up to the end of the 1980s. Special thanks go to the director of this archive, Mr. Thomai. I also wish to thank Ferrara University in Italy for the research period I spent there and for the feedback provided by my colleagues there.

Finally, I would like to express my gratitude to Polis University in Tirana, to which I dedicate this work, for fostering such a motivating research environment and for granting me time to write this book. It is Polis University that aroused in me the latencies that I never thought I possessed. I came to know about so many new things at this university, and I am highly thankful for the time I spent there.

Contents

1 Introduction 1
1.1 The Setting 1
1.2 Tirana as a Case Study 2
1.3 Methodology 7
1.4 Structure of the Book 8
References 11

2 The Natural City 13
2.1 The Natural City as an Emergent Phenomenon 14
2.2 The Natural City as Complementary Nuanced Variety 20
2.3 The Role of Architects and Urban Planners 22
2.4 Conclusions 27
References 28

3 Additional Theoretical Lenses to Understand the City 31
3.1 From a Broken Mirror to a More Holistic View 32
3.2 Understanding the City from the Quantum Paradigm's Holistic Reality 34
3.2.1 Emergent Reality and the Wave–Particle Duality 34
3.2.2 Multiple In-Between Gradients 36
3.3 Understanding the City from the Fractal Perspective 37
3.3.1 Relational Emergent Order 37
3.3.2 Principle of Scale, Repetitions, and Self-similarity 38
3.3.3 Form as a System Structure 39
3.4 Understanding the City from the Complexity Perspective 40
3.4.1 Self-organization and Emergent Realities 40
3.4.2 Urban Patterns as a Complex Behavior 41
3.5 Conclusions—Emergence as a Key Concept 42
References 43

4 Historical Notes on Tirana 45
4.1 Transformations and Events 46
4.1.1 The Emergence of the Organic City (1614) 46

4.1.2 King Zog and Stile Littorio 51
4.1.3 Framing the City 56
4.1.4 Spontaneous Waves of Urbanization 64
4.2 Conclusions 72
References 75

5 Tirana Patterns at a Glance 77
5.1 Patterns Representative Areas in Tirana 78
5.2 Pattern #1: Historical Organic 82
5.2.1 The Essence of the Pattern 82
5.2.2 Brief Description of the Representative Areas 86
5.2.3 The Simulacrum of the Historical Organic 89
5.3 Pattern #2: Recording Over 100
5.3.1 The Essence of the Pattern 100
5.3.2 The Main Elements of the Recording Over 107
5.3.3 The Representative Areas 110
5.4 Pattern #3: New Organic 116
5.4.1 The Two Species of New Organic 116
5.4.2 Compressed New Organic 119
5.4.3 The Representative Area 125
5.4.4 Inflated New Organic 126
5.4.5 Characters of the Inflated New Organic 132
5.4.6 The Representative Area 136
5.5 Conclusions 143
References 144

6 Holistic Approach to Tirana Pattern Analysis 147
6.1 Understanding Wholeness Through Relationships 148
6.1.1 The Unbroken Web of Relationships 148
6.1.2 The Relational Order and Dynamic System Structure 149
6.1.3 Relationships as Self-organized Behavior 150
6.2 Quantum Approach—Patterns as Overlaps Within the Unbroken Web 154
6.2.1 Historical Organic as a New Emergent Reality 154
6.2.2 Recording Over as a New Emergent Reality 166
6.2.3 New Organic as a New Emergent Reality 171
6.3 Fractal Approach—Patterns as Relational Order 185
6.3.1 Historical Organic 186
6.3.2 Recording Over 192
6.3.3 Compressed New Organic 194
6.3.4 Inflated New Organic 196
6.4 Complex Approach—Urban Patterns as the Human Capacity to Self-organize 201
6.5 Conclusions 208
References 210

7 Methodology for Holistic Understanding of the Urban Patterns 211
7.1 Measuring Patterns as Complex Behavior 212
7.2 Methodological Steps: From Observation and Analysis to Essential Quality and Information 213
7.3 Modeling That Reflects the Specific Urban Phenomena 221
7.4 Conclusions 224
References 225

8 Conclusions 227
8.1 Analytical Categories to Observe and Analyze Emergent Qualities 228
8.2 Looking at Patterns as Emergent Realities 229
8.3 Self-organization at the Base of the New Emergent Realities 230
8.4 The Role of the People in an Open System 231
8.5 Methodology Supporting Co-evolution—Impact on Urban Planning and Design Practice 231
8.6 Education—Greater Focus on Relationships than on Objects 233
8.7 Tirana and Beyond 234
References 236

Appendix A: The Measurement Process 237

Appendix B: Patterns Emerging from a Set of Rules 249

Chapter 1
Introduction

Abstract In this chapter, I provide information on the main purpose of the book, the case study, the methodology, the structure of the book, and the main ideas contained in each chapter. This book discusses new ways of looking at and understanding cities by applying methods and approaches based on the conceptual grounds of quantum, fractal, and complexity theories. Considering self-organization as the starting point at the base of complex systems, this book tries to understand how specific qualities emerge and evolve from this behavior. Then I present some important points as to why Tirana was chosen as the case study to examine notions of spontaneous self-organization and emergent urbanism. I focus on Tirana's specifics and attempt to understand urban phenomena through an inquiry process that involves a theoretical perspective on urban analysis, urban observation methodologies, and paradigms under which the observations are conducted. Further, I give a brief description of the methodology. The investigation used a mixed approach that combined qualitative and quantitative methods to investigate complex issues and achieve the main purpose of the book. As mentioned, although the theory in this book is illustrated using the specific reality of Tirana, the logic is valid for cities in general.

Keywords Cities · Self-organized · Quantum · Fractal · Complex · Tirana

1.1 The Setting

This book discusses new ways of looking at and understanding cities by applying methods and approaches based on the conceptual grounds of quantum, fractal, and complexity theories. Assuming the case study of Tirana, Albania, I reflect on how we think about cities, observe, analyze, and transform them. As cities evolve over time, they also develop unique characteristics reflected in their morphological and spatial characters, in the specific combination of activities, or in the networks of social interactions occurring within them. In Tirana, many people tend to solve their problems outside authority control and self-organize "into a collective whole that creates patterns, … evolves and learns" (Mitchell 2009, p 4). These concepts that belong to the science of complexity are the initial motivation to deal with the atypical Tirana urbanism, not only from the formal point of view but also from spatial and

S. Dhamo, *Understanding Emergent Urbanism*, The Urban Book Series,
https://doi.org/10.1007/978-3-030-82731-1_1

inter-relational perspectives. Considering self-organization as the starting point at the base of complex systems, this book tries to understand how specific qualities emerge and evolve from this behavior. Therefore, related to the new ways of looking at and understanding cities is understanding how complex systems interact and "exhibit emergence" (Holland 2014, p 27).

The lively atmosphere (Fig. 1.1) created by informal street vendors, the functional blur reflected in endless creative adaptations, the shifting meanings between public/private, illegal/legal, public spaces that are privately appropriated and private spaces that suddenly transform in public, or other phenomena occurring within the physical and human tissue of the city tend to be simplified, because they do not comply with our predisposition about how the world should be. Instead, the multiplicity of the same place, or "slipperiness," as Dovey (2012) calls it, due to the strong demand for the use of space, is also common ground for the "negotiable forms of governance and urban planning" (p 380) based on the understanding of underlying dynamics that guide these phenomena. These are typical cases wherein everyday rhizomatic practices clash with the hierarchical system of planning control (Dovey 2012, p 380), and this is my main concern.

In this respect, the ideas presented in this book are an effort to explore ways to consider specifics from a given reality and to regenerate and give them life in the planning and design process. The underlying patterns of these phenomena are elusive and may occur unseen or unobserved, unless we enlarge the focus of observation through additional theoretical lenses to gain a more holistic understanding of reality. This might be an alternative to the city suffocated by the constant exclusion of organic and natural phenomena. This book, therefore, seeks to outline a methodology for observing, analyzing, and potentially designing the city, with the resulting information being used as input for a model capable of reflecting the qualities of a specific reality.

1.2 Tirana as a Case Study

Before going further, I present here some important points as to why Tirana was chosen as the case study to examine notions of spontaneous self-organization and emergent urbanism. The traces of creative spontaneity can be seen everywhere in Tirana. Much of this city has been made by people, thus falling outside the control of authorities. This condition, which is reinforced by a wide variety of ubiquitous, enthusiastic, and amateurish architectonic languages, can be perceived at a glance as a form or incoherence to a lack of continuity in the urban environment. As will be seen further in this chapter and in Chap. 4, this is also due to the overlapping and interlocking of conflicting, suspended actions undertaken by authorities or citizens, often representing opposing interests. However, as I argue throughout this book, carefully observing this condition reveals a previously unknown world that helps interpret it in light of the diversity and richness of the urban text. For Tirana, this

Fig. 1.1 Market and residential area in Tirana, 2021 (Brari and Hasanaj)

kind of spontaneity at the city and neighborhood levels is presented in Figs. 1.1, 1.2, 1.3 and 1.4.

This antagonistic energy has played an important role in shaping and reshaping Tirana as a physical and ideal structure. Such energy has been an important driver moving the city between dreams and reality that, at times, may be difficult to manage.

Fig. 1.2 Panoramic transect of Tirana and periphery sprawl, 2021

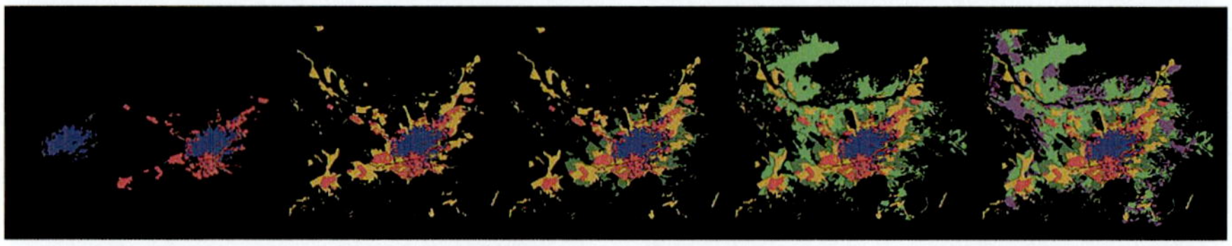

Fig. 1.3 Expansion of Tirana from the early 1920s to the first decade of the 2000s (reproduced from Aliaj et al. 2010)

Fig. 1.4 The birth and expansion of an informal neighborhood (reproduced from Aliaj et al. 2010)

Tirana is a city where the balance between authoritarian decisions and spontaneity has always shifted easily from one extreme to another; where authorities, perceiving it as the only way to move the city from spontaneity to organized urban planning, have overseen the exhaustion of human properties and the natural energy of the city; where users, during periods of weakened state power, have unofficially assumed control, creating dynamic and indelible transformations on the ground; and where the whole city became an arena for revenge on all that was considered public, as a form of social therapy to release the frustrations of authoritarianism. Yet, ironically, this city is also where the people have often been more sensitive to and protective of the city's sense of place than the power-aligned authorities and architects.

Tirana is a typical case in which informality has become the common practice to solve the housing problem under conditions, wherein the majority of the population is excluded from the legal practices to access affordable housing. A typical self-organized informal settlement in Tirana is not a slum-like settlement built with whatever temporary materials; on the contrary, the houses in these settlements are durable and permanent. However, from the beginning, enough flexibility is left to adapt in space and time to the growing needs of a family. The signs of this strategy are visible in the overlapping rebars or in the metallic armature, which is always ready to welcome additional floors or side extensions. This is a constantly changing building that holds the dream of an uncertain future and a never-ending neighborhood that transforms throughout the years via potential micro-adaptations. In this sense, informality in Tirana possesses specific, emergent characters and clearly expresses anthropological principles. As Di Raimo says about informality in Tirana, it shows inventiveness and a "certain anarchic character" that is recognizable in the expression of "many artsy and ungrammatical architecture elements," which are implanted in the built environment (2020, p 12). I present more details about this in Chaps. 4 and 5.

Hence, informality in Tirana is not only related to providing merely a shelter but also to the dream of a bigger and better house at a lower cost or to the dream of owning a business that would otherwise be unattainable with formal practices. In Tirana, it is difficult to find an unmodified building, both because the architects do not consider the needs of the occupants, and because they do not fully understand the model of life that generates that reality.

Tirana's experience is symptomatic of a poor planning system and a lack of trust in this system's capacity to solve real problems. The widespread perception is that Tirana represents the kind of planning that, in the name of rules, does not serve the public good. This is reflected in a clear antagonism: on the one hand, the ongoing struggle of people living in informal settlements for their acceptance, including negotiations with political forces and protests and, on the other hand, the always present threat for redevelopment and eviction by using police forces (the last case of eviction occurred in early 2020). This lack of trust, which is also deeply rooted in the Albanian mentality, generates and motivates the process of self-organization as a silent war that erodes and overthrows the system in moments of weakness. Such a phenomenon repeatedly happened in the period between 1991 and 1993, in 1997, and in 2005. It is expected to happen again, because no substantial measures have been taken to mitigate the housing problem.

With the exception of Law N. 9482 (03.04.2006) that—for the first time—acknowledged the need to formalize, urbanize, and integrate informal settlements, no serious steps have been taken, including the implementation of this Law. Between political and electoral games, the process remains unclear. This situation also further increased informality and deepened the lack of trust. While Tirana gained unprecedented physical and socioeconomic dimensions due to informal sprawl, there is no effort from governing structures to learn from the adaptive and resilient experiences of the informal settlements that, in many respects, are ahead of the urbanization processes. The population of Tirana before the collapse of dictatorship in 1991 was

250.000 inhabitants. According to INSTAT (2021), The National Institute of Statistics the population of Tirana County in January 2021 was 912.190 inhabitants. In this sense, informality has been an integral part of efforts to achieve social and economic progress in Tirana as a metropolis undergoing formation.

Tirana can be considered a laboratory to observe the correspondence between the complex structure of human interrelationships and the emerging morphological aspects of organic and informal settlements. As I try to show in Chaps. 4, 5 and 6, there exists a hidden world of self-organized interrelationships to be discovered—starting from the family level up to groups and affiliations at the neighborhood level. I also demonstrate how these groups filter the flow of information and ensure accepted social behavior; how these groups generate and recycle self-organizing energy, which results in governance arrangements; and how human principles, which trigger these relationships, are embodied in the morphological expressions of informal settlements, among others.

Last but not least, Tirana is a multireligious society. As I will present in Chap. 4, Albanians—for historic reasons—had to adapt between Christianity and Islam. The census by religious affiliation carried out in Albania before the end of the Second World War showed that 70% of the population were Muslim (divided into two main groups: Sunni and Shia), and the rest of the population were 20% Christian Orthodox and 10% Catholic (Della Rocca 1990, p 8). After the Second World War, the communist regime banned religious practices and beliefs via Decree N. 4337 (13.11.1967), which was later reinforced by Article 37 of the 1976 Albanian Constitution. After the collapse of the dictatorship in 1991, the religious issue reemerged, and religious communities began to recover. During the last three decades, Albanian people have never been keen on joining a new census related to religious affiliation. In fact, this was a widely debated topic among the public when a new population census was done in 2011. This vague attitude toward religion fostered uncertainty about actions and approaches to social life, while also creating more space to pragmatically and flexibly shift from one mentality to another. From this point of view, Tirana is a typical model wherein different religions—instead of producing separation and dividing cities—actually create the space of tolerance. This unique ambiguity or even multiplicity that Albanians adopted related to religious issues was reflected in neighborhoods and cities that readily absorbed people from different religions and cultures and were always open to unknown/undetermined things. This is true even in the cases of the organic parts of Tirana generated under clear principles of the Islamic city, which I describe in detail in Chaps. 5 and 6.

In conclusion, every city—in one way or another—carries traces of spontaneity and informality. However, in Tirana, informality is a unifying element that swept almost the entire city, with rules and regulations being violated everywhere. Informality and self-organization, as silent challenges to authoritarianism, are fueled by a lack of trust in government structures. As such, they simultaneously express repression, liberation, desperation, and euphoria (Di Raimo 2020, p 12). This tumultuous trajectory clearly indicates cyclical declines in spontaneity, which have blurred the rigidity of top-down planning decisions. The history of the city is punctuated by continual restarts, both by residents creating their own rules and by authorities fighting against the city. Such a tension is materialized in the collage of fragments containing the natural city's latent enzymes, which sit—ready to attack—should the authoritarian system's "immunity" suffer from declining control. These points, and many others that I present in the specific chapters, support the thesis that Tirana is an ideal model through which to study the self-organized emergent urbanism and the specific processes characterizing this model.

1.3 Methodology

In relation to the main purpose of this book, I investigate new ways of looking at and understanding the reality of our cities from a more holistic perspective. To achieve this goal, I explore new lenses or theoretical concepts that help present a wider range of phenomena without simplifying the complexity of the real world. These also facilitate the exploration and analysis of the correspondence between the complex structure of interrelationships or governance arrangements in a self-organized process, on the one hand, and the emergent qualities embodied in the morphological aspects, on the other hand. Furthermore, these concepts outline a methodology that can potentially be used for observation, analysis, and design purposes. The case study of Tirana addresses both the main purpose of the book and the exploration of the above issues. For this, I focus on Tirana's specifics and attempt to understand urban phenomena through an inquiry process that involves a theoretical perspective on urban analysis, urban observation methodologies, and paradigms under which the observations are conducted.

The investigation used a mixed approach that combined qualitative and quantitative methods to investigate complex issues and achieve the main purpose of the book. I started by investigating the urban phenomena qualitatively: after presenting a summary of additional theoretical lenses for a more holistic understanding of the city (Chap. 3), I analyzed whether the observed phenomena in Tirana met any theoretical preconception, after which I assessed and tuned the theoretical lens accordingly to evaluate a range of hidden qualities embodied in the urban environment (Chaps. 5 and 6). This phase, among others, involves conclusions related to theoretical concepts, which comprise the bases for observation and analytical processes, as well as analyses of the urban phenomena in Tirana and in the respective urban patterns, including descriptive labels for each of them. This qualitative phase provides evidence of additional knowledge from using the reframed conceptual basis and prepares the conditions for the application of the quantitative method, translating, and expressing those qualities—invisible and latent so far—in a set of regularities. This phase involves a detailed measurement process, statistical elaboration and interpretations, and the organization and visualization of the generated information through tables and diagrams (Chap. 7 and Appendix A). Both phases are crucial for the next stage, which aims to bring the resulting information back into a design process and use that as an input for a model that is capable of reflecting the qualities of a specific reality (Appendix B).

All the related data directly collected or surveyed on-site included the following: results of spatial and morphological analyses, including a typological survey to identify the range of structures characterizing each pattern, both at the city and area (neighborhood) levels; areal imageries and historic maps to investigate transformations of the urban form, distinctive characteristics in the spatial configuration of neighborhoods (being those formal/informal), and the distribution of these characteristics at the city and the larger scale; and archival research related to projects and visions about Tirana starting from the 1920s, reports and socioeconomic surveys from international or non-governmental organizations (NGOs), legal acts, and historical photographic research. In addition, the investigation involved an extensive mapping and diagraming process, which aimed to provide a visual sense of the main arguments and to illustrate the information and analytical content. The process also required countless field visits and photographic surveys at places where certain parts of the book were conceptualized and written on-site.

Finally, although the book focuses on the case study of Tirana, I try to express the analytical path in a more generalized methodology (Chap. 7) for observation and analyses. These can potentially lead to a planning and design process based on an understanding of specific urban phenomena from a more holistic perspective. This methodology can be used to understand, predict, and influence events and occurrences that are deemed chaotic beyond Tirana. Thus, in a more generalized framework, the proposed methodology helps to answer—in a different and more thorough way—the important question of whether the urban dynamics and transformation of our cities can be understood from the theoretical perspective of interactive relationships across spatial and temporal scales. From this perspective, this methodology embodies the idea of more flexible urban planning and design by utilizing pliable instruments that inform and promote the co-evolution of formal and informal parts of the city. It also encourages the growing together of the range of actors involved in the development, through learning and adapting in the process and by recognizing the need for "nuanced and contextual" governance (Suhartini and Jones 2019, p 220). This logic, as I argue throughout the book, is an important message that can be applied to any specific environment.

1.4 Structure of the Book

I start by presenting, in Chap. 2, the significant gap between the reality on the ground in cities and bureaucratic planning. To demonstrate this, I introduce the concept of the natural city and describe the characteristics such cities develop as a reality that respects human concerns at the individual or group level. The natural city arises through the creativity of ordinary people and embodies human properties in a genuine and pure form. Understanding this human aspect of natural cities is one of the main interests in relation to the purpose of this book.

Chapter 3 introduces three theoretical concepts to feed the urban planning and design field: the quantum paradigm, the fractal perspective, and complexity theory.

These concepts, which together allow a more holistic approach, underpin the observation and analytical techniques used to interpret the latencies and multiple realities of a world that, as Zohar and Marshal (1994, p 62) stated, is an "unbroken web of overlapping … internal relationships." It was crucial to detect and interpret these overlapping relationships early to understand, in a different way, urban wholeness and its successive transformations in new emergent realities. In the quantum world, these new emergent realities are not the sum of their parts, because interrelationships are involved in their creation.

In Chap. 4, I observe and analyze a specific reality: the urbanization of Tirana. Here, the natural city of individual arrangements coexists with fragments of highly centralized planning and authoritarian design, creating specific dynamics. Chapter 5 outlines the main characteristics and typical urban patterns of Tirana, which has experienced cyclical declines in spontaneity, mitigating against the imposition of top-down order. The recurrent mutual attacks between the organic, or informal and designed city became apparent. As a result of this interaction, whether violent or organic and regardless of the city's shape and geometry, the city became more natural over time. This helped also to repair the dichotomy between formal and informal.

Chapter 6 presents the application of the three main approaches through which I apply the concept of wholeness in the field of urban patterns (understood as self-organized entities with site-specific emergent qualities). First, I see urban patterns as local temporal realities, but as part of an unbroken continuum in time–space arising from relationships of present and previous elements, with material and/or non-material aspects. This logic is based on quantum theory's wave–particle duality, in which apparently separate things are "aspects of some larger whole" (Zohar and Marshal 1994, pp 58–59) because of the arousal effect on the local temporal environment of interfering patterns from different space(s) and time(s). This notion of the unbroken web also appears in Arida's (2002, p 157) concept of the "society–space–time (SST) continuum … as an energy field of potential events." This helps to conceptualize the correlations between events in the specific history of a city and the osmosis of different anthropological cultures, places, or times in a web of wholeness that exists beyond the local society–space–time. Based on this logic, urban patterns are not static entities but a dynamic succession of horizons or new emergent realities arising within the society–space–time continuum, interfering with each other and with the local temporal environment. Therefore, in this book, patterns are considered local temporal manifestations of horizons (territorialities) "caught" in the society–space–time web of wholeness.

The second approach used for exploring the wholeness of urban patterns comes from "fractal city" theory (Batty and Longley 1994). According to this approach, patterns contain, internally, an invisible underlying structure of relations. Thus, the degree of order apparent on the surface actually goes deeper and arises from relationships and a hierarchy contained in the internal form. The structured system manifested by these internal features has its own statics (an aggregation of elements in subassemblies, creating distinct hierarchies) and dynamics (recursive behavior at the base of which remains the repetitive nature of irregularities across scales). This relates to what we consider irregular, planned or unplanned, formal, informal or organic. In this sense, the concept of "organic" is understood as individual arrangements at the finest scale (Batty and Longley 1994, pp 8, 28, 31) and as part of an open system that tends to self-regulate, which runs counter to the prejudicial and simplistic interpretations that merely consider large parts of the city under the category of informal. As Pagliardini et al. (2010, pp 331–351) argue, we need to learn from "spontaneous geometry," human-scale solutions, and the morphology of living urban structures to embrace "self-organization as a key feature" of success in the urban space.

The third approach I use to explain and apply the concept of wholeness is drawn from "complexity" theory. First, I demonstrate that the urban patterning process is a complex phenomenon that involves "large networks of individual components" that enter into relationships through "signaling and information" exchange (Mitchell 2009, pp 12–13). As Batty (2013) explained, the new science of cities depends on defining these relations. Here, wholeness features qualities that arise only through the relationships and interactions of a multitude of individuals that develop from self-organization to self-regulation. Complexity theory helps to identify how the invisible and unbroken space–time network of relationships and the invisible relational order emerge from cyclical interactions within a self-organized collective whole. This cyclical generative process is synthesized in Chaps. 6 and 7 in the correlation factors of motive, behavior, and modality, which embody the analytical categories of the first and second above-outlined approaches and attribute human meaning and awareness to this process. These factors are evident in the forces interacting within the system (the collective whole), from individual behaviors to the social modality, and, most importantly, in the story of a new emergent reality.

An important issue taken up by this book is how to translate complex and emergent behavior into operational information. In relation to that, Chap. 7 outlines a methodology that uses the introduced theoretical concepts and that starts from observation and analysis to the understanding of essential quality and translating this quality into information. A set of analytical categories is also given as part of this methodology for the more holistic observation and analysis of urban patterns as emergent qualities. The approaches used in this book helped to frame urban patterns, among others as computational and algorithmic information. This is illustrated in Chap. 7 and Appendixes A and B, in which I analyze the recursive irregularities that characterize the essential quality of a selected sample area and translate this into information to serve as input for a modeling process using generation techniques.

Having overcome the initial challenge of extracting latent information from the city as it is, using additional theoretical lenses, a second challenge was how to organize and use this information (as shown in Appendix A) as input for a scientific model, formalized within a computational framework based on parametric relationships that reflect the essential qualities of a specific reality. This model can help to further understand the phenomena also to use as a design instrument in a specific reality. Appendix B presents experimental work in this direction; however, no definitive design solutions are given, to fit them within a self-organizing logic and to leave room for their coevolution with other formal and informal elements of the city. Rather than focusing on the model itself, I place more emphasis on the logical and methodological approaches used to arrive at the data necessary for a model that would allow a better understanding of reality, highlighting the dimensions that would otherwise be unobserved or inadequately considered.

As mentioned, although the theory in this book is illustrated using the specific reality of Tirana, the logic is valid for cities in general or any other site-specific conditions. This book aims to demonstrate the potential of a methodology that broadens the angle of observable reality through additional theoretical lenses. As will be shown, the paradigms under which the city is observed and the ability to include without prejudice the entire urban phenomenon are crucial for achieving a more holistic understanding of reality. Throughout this book, I reframe reality by "pushing" (storing) information from the field of the unconscious to that of the conscious. There is an important message in this logic: if our cities do not entirely fit within the rules based on mechanical concepts, rather than changing the city, we need to assess and modify our observation and analytical lenses accordingly to evaluate that specific environment.

References

Aliaj B, Dhamo S, Shutina D (2010) Between energy and the vacuum. Universiteti POLIS, Tirana, Co-plan

Arida A (2002) Quantum city. Routledge

Batty M (2013) The new science of cities. The MIT Press, Cambridge, Massachusetts, London

Batty M, Longley P (1994) Fractal Cities—a geometry of form and function. Academic Press, London, San Diego, New York, Boston, Sydney, Tokyo, Toronto

Dovey K (2012) Informal urbanism and complex adaptive assemblage. International development planning review

Holland JH (2014) Complexity—a very short introduction. Oxford University Press

INSTAT (2021) The Institute of the Statistics. http://www.instat.gov.al/al/temat/treguesit-demografik%C3%AB-dhe-social%C3%AB/popullsia/publikimet/2021/popullsia-eshqip%C3%ABris%C3%AB-1-janar-2021/

Mitchell M (2009) Complexity—a guided tour. Oxford University Press

Pagliardini P, Porta S, Salingaros N (2010) Geospatial analyses and living urban geometry. In: Jiang B, Angela XY (eds) Geospatial analyses and modeling of urban structure and dynamics, the GeoJurnal library, vol 99, Springer, pp 331–351

Di Raimo A (2020) What does informality have to say to architecture?—decolonizing the enquiry and the enquirer. In: Di Raimo A, Lehmann S, Melis A (eds) Introduction of part 1 in the book informality through sustainability—urban informality now. Routledge, London and New York

Della Rocca RM (1990) Nazione e religione in Albania (1920–1944). Il Mulino, Ricerca, Bologna

Suhartini N, Jones P (2019) Urban governance and informal settlements—lessons from the city of Jayapura, Indonesia. Springer, The Urban Book Series

Zohar D, Marshall I (1994) The quantum society—mind, physics, and a new social vision. Quill William Morrow, New York

Legislation

Brari & Hasanaj private photo collection. Tirana 2021
Decree N. 4337 dated 13.11.1967 On the suppression of religious confessions
Law N. 9482 dated 03.04.2006 On legalization, urbanization and integration of illegal constructions
Law N. 5506 dated 28.12.1976 Constitution of the socialist people's republic of Albania, article 37 declared Albania an atheist republic

Chapter 2
The Natural City

Abstract In this chapter, I first elaborate the notion of the natural city as a complex emergent phenomenon, spontaneously self-organized from the bottom, that adapts and evolves through coordination at the smallest and that represents a formal/informal continuity. In this regard, I explain the concepts of organic, informal, and sprawl cities, which are emergent and self-organized in their nature and highlight the fact that their characteristics are similar to those of the complex systems. Understanding their emergence as expression of human relationships that set in motion an internal universe, despite being formal/informal, serves the main purpose of the book about identifying new ways of looking at and understanding cities. In addition, the natural city, as an alternative to the artificial ones demonstrates the gap between the complex reality of cities and the simplistic approach of bureaucratic planning. Then, I try to link the topic of the natural city and the increasing trend of growth in informal settlements, with the role of the architects and urban planners and their positioning with regard to social housing and informality issues.

Keywords Natural city · Emergent · Organic · Informal · Self-organized · Human properties

In this chapter, I first elaborate the notion of the natural city as a complex emergent phenomenon, spontaneously self-organized from the bottom, that adapts and evolves through coordination at the smallest and that represents a formal/informal continuity. In this regard, I explain the concepts of organic, informal, and sprawl cities, which are emergent and self-organized in their nature and highlight the fact that their characteristics are similar to those of the complex systems. Understanding their emergence as an expression of human relationships that set in motion an internal universe, despite being formal/informal, serves the main purpose of the book about identifying new ways of looking at and understanding cities. In addition, the natural city, as an alternative to the artificial ones demonstrates the gap between the complex reality of cities and the simplistic approach of bureaucratic planning. Awareness of the complexity underlying the natural city encourages the improvement of the theoretical and conceptual basis for understanding the urban phenomenon.

Then, I try to link the topic of the natural city and the increasing trend of growth in informal settlements, with the role of the architects and urban planners and their

S. Dhamo, *Understanding Emergent Urbanism*, The Urban Book Series,
https://doi.org/10.1007/978-3-030-82731-1_2

positioning with regard to social housing and informality issues. In relation to this, I argue that the role and sensitivity of architects and planners to appropriate these issues as part of the discipline has changed over time. Currently, there is a need for architects and urban planners to learn how to include adaptive behavior and evolutionary dynamics in a more flexible planning alternative, which operates in the framework of a co-evolutionary logic. Finally, the need for change in the role of architects and planners relates to the necessity of inclusion in their educational curricula topics related to the understanding of informality and complexity issues.

2.1 The Natural City as an Emergent Phenomenon

The natural city with its formal, informal characteristics is a reality on the ground. This is the starting point for writing this book. As I argue further in this chapter and analyze throughout this book, what is here termed the "natural city" includes an emergent spatial order, generated by individual arrangements at the small scale, that respects human concerns at the individual or group level. Despite generation at the smallest, it has an impact on larger spatial scales by presenting a link between the underlying dynamics in how cities grow through scales and micro-adaptations that produce effects based on interactive cooperation. From this preliminary definition, the natural city is a self-generated urbanism that emerges without the help of any central controller, based on the capacity to self-organize, something often considered simply as informal. In fact, these characteristics, which I will return to several times in this book, are properties of complex systems. Over the past decade, the theory of complexity has grown in appreciation in order to investigate issues related to urban territories and processes, especially when it comes to informal and formal/informal relationships.

As many authors cited in the book (Dovey, Jones, Roy and AlSayyad, Silva, Di Raimo et al,, Kamalipour, and others), who treat the subject of informality from a more scientific and realistic perspective state, the term informal that quite often implies illegal, poverty, social marginalization, slums, squatting, or other not positive connotations, does not include the full complexity of the phenomena. As Silva argues, complexity theory helped to accept that a significant part of contemporary urban territories is unplanned and the result of self-organized processes leading to evolution emerged from bottom-up processes, as citizens' adaptive response (2016, p. 2), as an interaction between subsystems, or between "actors and their system". In this regard, if we accept that plans could be part of the system, they could include local initiatives and their evolution (2016, p. 10).

As I will present further in this chapter formal and informal are not dichotomous. Informality as a long-standing global phenomenon is intertwined with the so-called formal parts of the city. As Dovey says, even in the cases of cities that seem to have a formal character, formal and informal processes are mixed (2012, p. 372). These informal parts emerge through generative processes and grow through self-organized incremental adaptations (Jones 2020, p. 181). As Jones continues, they manifest through small-scale changes often considered not important since they are chaotic and lack a formal order contributing to the visual incoherence, multiplicity of functions and of morphological forms (2020, pp. 182, 190). This way informal settlements have been evolving as "meaningful places and communities … based on mutual interdependence" founded on tradition of underlying socio-cultural orders (Jones in Suhartini and Jones 2019, p. 20). This complex intertwining of formal/informal parts and processes represents the natural city.

In addition, and important for the ideas presented in this book, as Di Raimo argues, informality can bridge problems of contemporary cities as complex places growing through social transactions or contracts and architectural design that still remains at the scale of individuals or groups of individuals (2020, p. 7). Within the architectural dimension, informality is regarded as the transformative practice of the individuals through a set of rules beyond the control of the discipline of architecture (Di Raimo et al. 2020, p. 3). In fact, this process generates "something that should be acknowledged as an unconscious interpretation of architecture" (Di Raimo 2020, p. 18).

Roy and AlSayyad (2004, in Roy 2005, p. 148) also refuse to use the notion of informal and oppose the idea of a dichotomy between formal and informal. For them, informality is "a mode of urbanization" indicating "an organizing logic, a system of norms" governing urban transformations when a high demand for housing is left unaddressed. Informality is "a series of transactions that connect" economies and spaces, which are not separate sectors. In this sense, informal or unplanned is nothing more than "a state of exception from the formal order or urbanization" (Roy 2005, p. 147) or a manner of urbanization. There are some criteria traditionally used to categorize informal settlements. These include physical features that comprise the repetition of "non-geometric and non-linear patterns" (Suhartini and Jones 2019, pp. 20, 21), legal status of land, socio-economic quality, and the process of establishment. In addition, the urban forms and structures of informal settlements express resilience, sustainability, and affordable methods in the provision of housing and services, as well as "social and physical processes" that emerge on site, develop and transform the informal settlements (2019, pp. 20, 21).

The same term "natural city" is used by Alexander (1965, p. 1) in its seminal paper "the city is not a tree". For him, "natural cities" are those "which have arisen more or less spontaneously over many, many years," implying with this expression an increasing level of complexity over time. While the "many, many years" implies a long history of human processes and physical stratification, the conceptualization of natural cities in this book include also settlements generated over a short period; that is, spontaneous or informal urbanization. This idea is elaborated on later in the chapter, but first, we need to better understand the wide array of natural city expressions.

In this regard, it is useful to present here some similar phenomena which are related to historic processes of informality. Such informality is recognized and accepted as spontaneity with relation to the historical city, such as the organic growth of cities during the Middle Ages, when cities were still small and compact. As Dovey says, the medieval city morphology is informally produced by micro-adaptations over time

(2012, p. 373). What we admire as tourists in many European cities are exactly the fragments of informal settlements of the past, that going on the "safe" we call them organic or vernacular. Although the organic city and the informal city reflect nuances in the processes behind their emergence, particularly related to time stratification, both are similarly self-organized in nature. This similarity in the self-generation and self-organization processes from the bottom helps to better understand the emergent nature of the informal city. It is, therefore, important to first understand the organic cities and the network of relationships at the base of their emergent nature, which I will discuss in the following paragraphs and in Chaps. 5 and 6.

Authors such as Kostov (2003) and Batty and Longley (1994) have used the concept of the organic city. Kostov described this kind of city as "chance-grown," "generated," and "geomorphic" (Kostov 2003, p. 43). These notions are also the starting point for Batty and Longley (1994, p. 28), who took a fractal perspective of organic cities (a view further discussed in Chap. 6). But how can we understand "organic" in relation to cities? The Merriam-Webster Dictionary, among others, attributes to "organic" the meaning of something "derived from living organisms." In our context, this can be understood as something created by humans and as something strongly influenced by topographic features and other natural and human factors. Further, "organic" signifies the "systematic coordination of parts: organized as an organic whole." With regard to cities, this implies a clear and repetitive relationship between the elements constituting the whole; this will be crucial when considering the phenomena from the quantum, fractal, and complexity perspectives. Finally, the Lexico UK Dictionary defines "organic" as "characterized by gradual or natural development," which implies a series of processes and steps to be discovered, or its relationship within a time–space framework. As I will present in Chaps. 4 and 5, many parts of Tirana still present characteristics of organic informality, implying with this a longer historical stratification.

At this point, it is important to explain what kinds of forces are at work in the generation process of the organic cities and what relationships exist between the "urban form and the urban processes" (Kostov 2003, p. 9), which combines factors with social and physical nature. In the first group of factors, Kostov (2003, p. 62) includes "urban improvisations based on social structure and the limits of public control," which influence the irregular city form. As shown later in this book, social structures and community ties are important aspects of human behavior and underlie the urban patterns related to such behavior. Regarding the second group of factors Kostov (2003, p. 62) related these to the physical structure of the city and its change over time, as well as to a series of "earth-bound" factors (e.g., topographic conditions and land division) that play an important role in weaving the city into its geographic container, especially during the initial stages. Both of these aspects, which in this book will be labeled as "material" and "non-material" aspects, are important for analyzing the natural city.

Emerging from bottom-up spontaneous processes, mostly generated from the daily activities of their creators, the set of features and geometric qualities characterizing the organic city is fundamentally different from that of a designed city. This is not only due to the sense of randomness that characterizes the physical structure of the organic city—for example, street segments that unexpectedly refract and bifurcate based on pedestrian logic or crossings without an easily detectable pattern—but also to the sense of flexible and sensitive adaptation to the patterns of life. These characteristics, as Batty and Longley (1994, p. 35) stated, are reflected in the organic city's "informality, its idiosyncrasies and its picturesque properties" but not in the exhibition of standard geometries. The irregularity of their form in Euclidean terms, due to the absence of an overall controlling authority or designer, is often considered to reduce the acceptability of the organic city because of the human obsession with visual geometric order.

This does not mean that the organic city is uncoordinated; it just coordinates differently. The series of actions and processes at the individual level that create a continuous flux of variations does not fit with the categories of "planned" and "unplanned" (Kostov 2003, p. 51), which tend to oversimplify by not considering the complex processes that generate organic cities. As already noted, "organic" inherently implies a coordination of parts or relationships as an integral whole. Related to this issue, Batty and Longley preferred to stress that although organic cities do not show signs of planned geometry, they result from many "individual decisions which have been coordinated in the small." Because of that, they are more reflective of natural properties and concerns at the local level (1994, pp. 8, 28, 31).

Organic city embodies the relation between formal and informal in the historical sense of which comes first. As Dovey says (2012, pp. 373–374), it is generally thought that "informality precedes formality," such as the medieval nuclei in many European cities. However, there is also quite a contrary case when the formal city comes first and informality infiltrates within formal frames. This process is especially pronounced when the formal building structures were imposed physically and culturally to cancel a previous organic order. In these cases, once the geometric order has been introduced, it assumes the role of the auxiliary bearing structures for upcoming deregulation events, which normally stem from changes that have to do with specific conditions (e.g., socio-economic and political). Even in these cases, during their daily activities, people unceasingly adapt and redefine their livable space. Such minor but steady and gradual finetuning of internal and external space results in substantial or even complete alterations over the long term. This process is particularly evident in post-restrictive cultures, where small changes in equilibrium tip the balance against the controlling authority in favor of social improvisations in public spaces. These improvisations can be seen growing out of the artificially inserted geometric structures.

One example of such a relationship, which reshaped the geometry of previously built structures, is the effect of urban deregulation on Tirana's Communist-era housing blocks. As discussed in Chaps. 5 and 6, this process transformed these sterilized outer-related spaces into intricate ones as they were filled in and recorded over by parasitic structures. This created continuity in the urban fabric and helped to heal the wounds of dichotomous detachment inflicted by the modernist city. The basis of this transformation was the abrupt shift from imposed collectivism to individualism or "tribal" culture after the collapse of the dictatorship in 1991, which brought with it the deregulation of rigid geometrically designed space. These self-organized deregulated processes mean that "planned" and "organic"/informal may intimately coexist and, in the long term, be interlocked in an osmotic continuity of the natural city.

In the cases described so far, there is a component of informal spontaneity. In relation to the historic interpretation of informality, from the industrial era, and especially during the early nineteenth century, the view on informality started to change, with informality acquiring a negative connotation in the present day. Now, informality is associated with the sprawling city peripheries of the developing world, caused by the poorest social strata informally constructing affordable housing. Thus, urban sprawl as a general and specific concept is another important phenomenon related to the issue under discussion. While the phenomenon can occur formally or informally, it mostly depends on the availability of land and public control. Generally speaking, there is a negative conception of sprawl globally as "chaos, [and a] lack of structure or demonstrable catalysts" (De Geyter Architects 2002, pp. 21, 23). However, there are many specific expressions of urban sprawl, and it displays different forms in different areas of the world. For example, as shown by Malusardi and Scoppetta (2008, pp. 2–3), "urban sprawl" in the US is related to urban diffusion and correlates with the "culture of colonization" and the immense territory to be occupied; the colonization of Africa followed a similar pattern. In Europe, the situation was different: the limited availability of land made planned development necessary. In Mediterranean countries, the phenomenon was mostly represented by "illegal house building produced by squatters" on the peripheries of large cities, such as "Istanbul, Ankara, Tirana, Roma, Cairo, and in the large cities of North Africa" (Malusardi and Scoppetta 2008, pp. 2–3). This same process occurred worldwide. In Western Europe, increasing urban sprawl was characterized by widespread diffuse urbanization around six of the most developed areas, including metropolitan London in England, the Randstad in the Netherlands, the Flemish Diamond in Belgium, the Ruhrgebiet in Germany, the Border Triangle of Basle-Zurich-Bern in Switzerland, and the Veneto Region in Italy (De Geyter Architects 2002, pp. 14, 33). Urban sprawl, according to Malusardi and Scoppetta (2008, p. 3), is the "spontaneous illegal growth of the periphery," unfolding steadily, slowly, and almost inevitably. This manner of growth, which might be "unsustainable but alive and real" flourishes under conditions of low public control.

As mentioned at the beginning of this chapter, there are important advances in understanding the phenomena of urban informality from a new theoretical perspective. In this regard, Dove proposes to re-think the dynamism of urban change and informal/formal conception through the concept of complex adaptive assemblage (2012, p. 371) using two complementary theoretical frameworks: assemblage theory referring to Deleuze and Guattari where the focus is on multi-scalar interconnectivity at the level of building, street, neighbourhood, district and city (2012, p. 379) and flows between parts generating the dynamism between formal/informal (2012, p. 376); and the theory on Complex Adaptive Systems (2012, p. 371), "where the behaviour of the system depends on unpredictable iterations between parts." As I will explain in Chap. 6, "the outcomes of such systems cannot be determined … but rather emerge from … adaptation and self organization." This emergent reality is a mix of formal/informal properties and practices (2012, p. 377). Thus, as this author argues, "Informal urbanism is a mode of production of urban space" (2012, p. 386) assembled through its multi-scale connections (2012, p. 380). In this sense, complexity constitutes the perfect scientific environment for a "renewed tradition of urban and architectural observation" in which formal and informal are intertwined with agents of transformation (Di Raimo 2020, p. 17).

While the terms "natural," "organic," "informal," and "sprawl" overlap in their usage in this book, they are not fully interchangeable because of nuances due to the distinct historical and socioeconomic contexts from which they emerged. As already mentioned, while the term "organic," alongside its connotations of spontaneity, often indicates a historical anthropological stratification, the term "informal" is directly associated with the stigma of law violation and the phenomenon tends to be excluded from the architectural and urban domains from the outset. Conversely, while both these processes start and continue spontaneously, "informal" areas carry different qualitative and quantitative social and spatial problems, mostly due to the time pressures and short duration of the moment from which they emerge. This does not mean that their morphologic and cultural components need to be blindly annihilated due to their perception as chaotic, without structure, and socially faceless.

The inclusive notion of the natural city, as used in this book, embodies the characteristics described by Alexander (1965, pp. 1, 2, 5), who viewed the natural city as an "abstract structure," a "collection of many small systems" that make up a complex system, and a structure of "sets" that cooperate under certain conditions of varying complexity (e.g., the "semilattice" or "tree"). For Alexander, the natural city is an alternative to the "artificial cities" created by designers, which he considered unsuccessful from a human point of view (1965, p. 1). The natural city also includes the "organic" characteristics described by Kostov (2003) and Batty and Longley (1994), which, alongside historical and cultural stratifications, include self-organized city-shaping processes that reflect the social, cultural, and community structures as an important conductive network of bottom-up information exchange. Most importantly, the natural city includes the characteristics of the so-called "informal" city, which implies a rapid, spontaneous, and abrupt urbanization process created under a series of constraints, including time pressure, risk of prosecution and/or persecution from authorities, and social prejudices and frustrations. This kind of city does not necessarily embody physical historical stratifications, but expresses "a mode" indicating "an organizing logic," which is a rich and complex mental and cultural imprint (Roy 2005, p. 148). The natural city includes the general characteristics of "urban sprawl," with its lively spontaneity and apparently random and chaotic nature. The natural city is a mode of production of urban space; it micro-adapts and is a multi-scalar assemblage (Dovey, p. 380) that evolves through self-organized adaptive response of citizens (Silva 2016, p. 2) or interactions between parts and the system. In this sense, the natural city, includes the characteristics of informal settlements in relation to coevolution within the framework of groups and individuals that self-organize across space, time and society (Jones 2020, p. 181). These incremental micro-adaptations interlock formal and informal in a continuous whole.

In this regard, the natural city is a way to repair the dichotomy created by rigid planning and to overcome the psychological gap between the city made by people and the one created by designers. The natural city is an input to include the less-developed parts of the world in which much of the urban development of the twenty-first century is taking place. Roy (2005, p. 147) highlighted this paradox in research on cities: while most of the "theories of how cities function remain rooted in the developed world," we need studies that match "First World 'models' and Third World 'problems'." The natural city can help to blur this dichotomy.

Finally, the notion of a natural city summarizes the main characteristics of self-generated spontaneous processes (independent of their legal status), which based on a motive (e.g., cultural, historical, or survival), puts into motion a social structure to settle and self-organize from the bottom up, through information exchange and coordination at the smallest or individual level, and to gradually evolve and adapt flexibly and sensitively to the rhythms of life. This occurs without central control and with an apparent randomness to its physical structure. As mentioned at the beginning, these are characteristics of complex emergent systems.

2.2 The Natural City as Complementary Nuanced Variety

As will be addressed in later chapters, in observing Tirana's urban condition as a natural city, there remains a degree of ambiguity and uncertainty regarding the classification of some fundamental categories, such as public–private, designed–non-designed, and even functional uses and relationships. Depending on the specific circumstances, each of these categories can simultaneously contain a nuanced variety of in-between statuses.

For example, the category of public–private, whether planned or unplanned, may contain a variety of temporary and transitional statuses depending on the actions that individuals or community groups undertake in their mutual testing process, which can vary according to the circumstances. During this mute negotiation process, carried out at the smallest scale, parties test permissibility to approach (or to touch) one or the other "extreme" statuses of a category, guided by perception. After each step they undertake, their actions may vary between abrupt invasion, imposition, gradual mediation, or interference. After such actions, a category may become another in-between state, exploring and exploiting the full potential contained "within" the category. This chain of actions and consecutive changes creates a blurred and interlocked situation that triggers infinitely further individual and/or group negotiations. The key point here is that there is a specific social mentality that consciously blurs reality to expand the space between two corresponding categories, which is filled with nuanced variations, creating a wider range of options. As such, it can be challenging to determine the urban condition of natural cities; the degree of uncertainty makes any effort to achieve this akin to playing a never-ending game. A similar process is seen in the many uses of a city, which blur the boundaries between regular–irregular or community–individuals (Fig. 2.1). This is because, in this kind of city, sociocultural factors (e.g.,

Fig. 2.1 In-between categories: actions test the use of space and transform it into a complementary variation; informal appropriation of streets and public spaces

obedience to rules, the ability to solve problems through informal arrangements, and a lack of trust in formal procedures) act as a hidden frame, intimately embodied in the neighborhood structure, outweighing urban regulations.

In the following chapters, I argue that a more holistic approach is needed to understand the reality of the natural city. In fact, the extreme statuses of a category, and the nuanced variation between them, are more complementary than mutually exclusive. In the next chapter, I explain this idea from the perspectives of quantum "indeterminacy" and "nuanced complementary," to shift from an "either/or" to a "both/and" way of thinking (Arida 2002, p. 142).

Clearly, isolating the main characteristics of the natural city presents some difficulties; the shifting of the categories through different statues creates a high level of complexity. This complexity clashes with the standard practices underpinned by a simplistic legislative approach. Here lies the great misconception behind the divergence between our perception of reality on the one hand and the role of statutory practices on the other: the deterministic approach to planning excludes natural phenomena by considering people empty and characterless entities.

In the present day, we are becoming aware that the naturally or organically growing city is optimal in many ways. As Alexander et al. (1977) observed, most of the wonderful places of the world were created by people; unless people "share a common pattern language" (1977, p. x), buildings and towns cannot come to life. However, authorities often reject urban conditions arising from humankind's creative interpretations, contributing to severe human interactions and the energy of cities. This discrepancy between the man-made city and order as understood in mechanistic logic is common in many cities around the world. Speaking about the increasing disconnection of people from their environment in metropolitan communities, Mumford (1961, p. 546) commented that everything in these ordered cities "is designed to kill spontaneity and self-direction," creating "a world they never made … ever more empty of human meaning."

2.3 The Role of Architects and Urban Planners

Many cities in developing nations lack the capacity to meet the demand for housing from the rapidly increasing urban population. At the global level, there is an increasing trend of growth in informal settlements, with "approximately 25% of the world's population in 2015 living in informal settlements …" approximating to one billion dwellers, the population which is expected to double in 2030 (Jones 2019, p. 1). This urban condition raises some concerns for the analysis that will be carried out in this book: this condition needs to be accepted as part of the urban reality, and the methodology used must be able to identify and include "irregularities" in the analysis without prejudice. Central administrations often propose urban development projects based on a simplistic standard geometric order and believe that changing a city physically can eradicate problems associated with human cultures, especially when these concern the urban poor. However, this offers only a temporary triumph. It would thus be of interest to see the positioning of architects and planners in relation to the public interest, particularly with regard to social housing and informality issues, and how this positioning has evolved and transformed over time. This is closely related to the debate on the role of architecture and urban planning today.

Before the Second World War, the modern architectural movement, especially *Congrès Internationaux d'Architecture Moderne* (CIAM), was characterized by an ethical interest in improving housing in cities in which an enormous number of people were living in miserable conditions because of problems inherited from the period of industrialization. This was an important moment; architects that had until this point stayed away from social problems now engaged in courageous experimentation to design low-cost collective housing. Based on a minimum social norm that ostensibly guaranteed a decent lifestyle for all, they tried to offer innovative typologies detached from the historical city, which was considered the origin of the problem. This detachment that led to the dissolution of the traditional city form, changed our perception about the city which was gradually transforming from a natural/organic phenomenon to an artificial one. According to Rowe and Koetter, the architects of this period seemed to have found the universal laws governing all nature. They legitimated their mission for a better society through a strong conviction that combined a firm belief in scientific truth and the humanistic myth (1978, p. 13). However, this dream gradually declined during the post-war reconstruction period, when many cities reached unprecedented dimensions.

Under the increasing pressure of post-war urbanization, some European governments took important steps to deal with the housing shortage, creating public institutions such as the *Istituto Nazionale Abitazioni* in Italy and *Habitation' Loyer Modéré* in France as part of state interventions to build low-cost housing. Although many new housing blocks were built as a materialization of the awareness about the social role of architecture, this activity was mainly based on mass housing production and heavily corrupted the principles of the modern movement, contributing to the uniformity of the urban environment and the ensuing alienation. A typical example is the *grands ensembles* built during the 1950s to 1970s in France. They were characterized by an urbanism of towers and "dams" which ended up in a deep social crisis from the 1980s.

During the post–Second World War period, the urban population rapidly grew because of continuous rural–urban migration. The housing situation at the beginning of this period was similar across the European capitals and can be clearly seen with reference to Rome, Italy, in scenes from the neorealist movies *Il Tetto* (The roof)[1] and *Le Mani sulla Citta* (Hands over the City).[2] In one such scene from 1950, a poor married couple illegally builds a one-room squat on the outskirts of Rome, taking advantage of the fact that, according to the law, if a building has a roof, the police cannot pull it down. The second movie tells the story of a cruel land developer in Napoli who uses political power for personal profit in a large-scale real-estate development. Typically, and of relevance to our discussion, new residential areas in the city periphery had to be built by large developers and/or architect-developers, who, as a result, gradually became key figures during this period and took precedence over architects.

In Europe during this period, the situation also varied between countries based on many factors. For example, the problem of informal settlements in Greece according to Potsiou and Ioannidis started in 1922 with massive migration "from Asia Minor into the mainland of Greece" (2006, p. 5). It increased after the Second World War until the establishment of the military dictatorship in 1967, when this phenomenon was significantly reduced only to reappear again after political changes in 1974. While the "first generation of informal settlements" created broad areas around

[1] *The Roof* (1956), Directed by Vittorio De Sica, written by Cesare Zavattini.

[2] *Hands over the City* (1963), Directed by Francesco Rossi, written by Francesco Rossi and Raffaele La Capria.

Athens and Thessaloniki (2006, p. 5), later generations emerged in vacation areas often used as second residences (2006, p. 8).

In the case of Lisbon, according to Ascensao (2010, p. 6), "slums" have been part of the city outskirts since the mid-1970s and 1980s. In the 1980s, over ten percent of the metropolitan's area population resided in informal settlements (Silva 2018, p. 2). Based on an evaluation of planning tools created over the past 40 years to formalize informal settlement in Lisbon Metropolitan Area, and how informal and formal systems impact each other, Silva speaks about a "healthier relation between formal and informal urban systems," not ignoring or controlling each other but instead learning with each other, or even "contaminating" each other. For this urban and building rules moved from temporary and exceptional adaption to specific needs of informal settlements, to permanent rules (2018, p. 3). In this regard, plans as part of the system can include the evolution of bottom-up initiatives (2016, p. 10) and this would allow co-evolution. As Silva Says, in the Portugal case, there are evident signs of co-evolution between informal settlements and planning institutions. These processes of co-evolution together with some examples from southern Europe (Italy, Spain, Greece) are considered to be innovative by United Nations (Silva 2018, p. 4). Under this evolutionary paradigm, the designer or planner "becomes a discreet part of the process" (2016, p. 3). As Marshal says, under this paradigm, cities are still somewhat organic complex and emergent (Marshall 2009, p. 262).

In the decades after post-war reconstruction, the housing crisis was transferred even to the developing world. Informal housing on city peripheries exploded in Latin American countries, especially where the phenomenon already existed, such as in Lima, Peru; Sao Paolo and Rio, Brazil; and Mexico City. The same methods and housing block typologies used in Europe were applied and again proved unaffordable and insufficient to deal with the scale of the housing deficit in these countries (Anklesaria 2013, p. 177).

During the late 1950s and mid-1960s, Peru became a prominent center for reflection on housing policies and the role of community development in self-organized help. However, only after the 1960s did architects accept the until-now disdained reality and consider the subject as part of the discipline. The British architect John Turner, who wrote about the informal settlements in Lima and their organization as communities, was influential in introducing the topic into urban and architectural discourse (Ballegooijen and Rocco 2013, p. 1). Convinced that people are best placed to understand their own situations, Turner advocated people being given the chance to manage their own housing issues and the concomitant withdrawal of the state from providing centralized solutions (Anklesaria 2013, p. 177). Turner emphasized community practice and mastery over the technical abilities of architects, engineers, and the state, instead of viewing the latter group as facilitators (2013, p. 178). This is reflected in his ideas about "self-help/community driven housing" (2013, p. 177) with minimal state intervention, which, combined with Hernando De Soto's arguments regarding the need to bring the informal economy into the mainstream market economy, greatly affected neoliberal agendas and the policy of international organizations such as the World Bank between the 1960s and 1980s. Indeed, self-help became an alternative to public housing production in South Asia and Latin America

(2013, p. 177). As such, there is much the world could learn from these dynamic and adaptive realities, which until now have been largely neglected. In fact, informal urbanism, is integrated "economically, socially, environmentally and aesthetically" to cities of the global South (Dovey 2012, p. 371).

Notably, in the framework of the neoliberal agenda, the architectural discourse regarding housing issues shifted toward self-help policies. This is reflected in the shift from the "messianic" role (Rowe and Koetter 1978, p. 13) of the architect in the modern era, as a protagonist in mass housing production, to the architect as enabler, responsible for helping to integrate their professional expertise with community knowledge and local resources.

In some Balkan and other former Communist-bloc countries, including Albania, such participatory processes arrived much later, either after the 1990s or at the beginning of the 2000s, coinciding with the collapse of dictatorships in these countries. Initially, the focus was on upgrading the informal urban peripheries created by massive urbanization processes under specific socio-economic conditions. These projects were introduced by pioneering local organizations, such as Co-plan active in Albania and in other south-East Balkan countries, who experimented with and introduced new methodologies for urban planning, as well as community cooperation and self-help housing under specific conditions.

Thus, returning to the role of the architect and the position of the discipline in relation to this issue, Aquilino (2011, p. 008) raised concerns about the worldwide "indifference" and "irrelevance" of architects and planners so far toward the most vulnerable parts of society. The fact that a large portion of urban growth is driven by non-formal architects highlights the separation of the profession from reality and its inability to contribute to solving real problems. Related to this issue, Batty and Longley (1994, p. 31) expressed their concern that despite more than 95% of all past and then-present cities being more organic in their form, the education of city planners was dominated by the geometric model and planning at the city scale. This imbalance between reality and the educational ideal has since become even more pronounced considering that most new urban transformations are taking place in the developing world, while urban informality is "seen as a generalized mode of metropolitan urbanization" (Roy 2005, p. 147) rather than a phenomenon exclusive to the peripheries of third-world cities.

This begs the questions: what needs to change in the approach of architecture and planning in relation to issues with high public impact, such as housing and urbanization processes? Should architects and urban planners return to being deterministic heroes, predicting end results and mechanically generating typologies alien to the historical and cultural backgrounds of communities? Should they remain as enablers, facilitating urban upgrade projects through citizen participation? Clearly, the classic participatory processes of the 1960s to 1980s, focused on community organization and giving people a voice, mostly or exclusively in poorer areas, are no longer sufficient. For this architects and urban planners should learn how to better combine statutory practices with participatory components and the "natural" processes of bottom-up self-organization; architects and urban planners should learn how to include adaptive behaviour within the urban planning and design practices in such a way that could lead to "co-evolutionary process between plans and society" (Silva 2016, p. 4), or "between planning institutions and informal settlements" (Silva 2018, p. 4); to encompass adaptions and evolutionary dynamics within the framework of a "tactical urbanism" (Silva 2016, p. 9) as a more flexible alternative to formal spatial planning.

Architects and urban planners should learn how to incorporate within the Spatial and design thinking the "assemblage thinking" as a non-linear logic according to which "the order of the city emerges unpredictably from the multiplicity" (Dovey 2012, p. 387); As Dovey says the great challenge of the urban planning profession, urban design, and architecture is "to move towards a model that accepts unpredictability" (2012, p. 387) and rethink the informal/formal conception. This challenge is "multi-disciplinary and multi-scale" for all cities of the global South, North, East and West" (2012, p. 374).

At this point, one thing is clear: we are still not at the stage to accept and consider the full spectrum of reality. As Jones says, informal urbanism and informal settlement as a global phenomenon are "underestimated as a major … issue in mainstream planning pedagogy" (2019, p. 14). Most importantly, he argues about the need to include in planning education an "understanding of the growth and complexity" (2019, p. 1) of the city; to expose students to the city shaping from the bottom resulting from specific combinations of "local rules, knowledge, and governance, in conjunction with the socio-cultural norms underpinning the dynamics of the broader" context (2019, p. 7); to expose students in a cross-cultural learning environment through international planning studios (2019, p. 14) as a key tool for "student-centered problem-based learning," where to experience informal settlements and informal urbanism as major global and sustainable development issues (2019, p. 3). These kinds of experiences make students questioning "the relevance of the conventional planning and urban design" (2019, p. 11). Kamalipour and Peimani also explore the ways in which urban design education can respond to informal urbanism by expanding its scope and including ordinary environments of informality (2019, p. 5). Considering the fieldwork as an integral part of urban design education they argue about the need to explore forms of informality (2019, p. 10) focusing on micro-scale self-organized process of creation and incremental adaption of space, however using a multi-scalar thinking (2019, p. 8). They consider education as an initial step towards the Sustainable Development Goal 11, in order to make cities resilient and inclusive (2019, p. 1).

Approaching the above arguments helps to reframe stigmas related to informal settlements; instead of drying up the human energies of the natural city in the name of rules with legal status, we can look at it as a more complex form of order in which individual choice is still possible. For this architects and urban planners should go deeper in understanding the enzymes or emergent qualities that make the natural city successful and use this information to inform a conscious design process. This methodology requires skills to understand the complex processes emerging from self-organization. In the next chapter, I will introduce additional theoretical lenses based on quantum, fractal, and complexity theory that can help to reframe our mindset related to thus far excluded realities. This methodology challenges the core of the urban design and planning disciplines, involves urban analysis, urban observation methods, observation paradigms, and contemporary debate on the role of urbanism and architecture in shaping the city.

2.4 Conclusions

In this chapter, I first elaborated on the notion of the natural city as formal/informal intertwining, organically emerged, self-organized from the bottom, and that potentially embodies dynamics that may lead to co-evolution with other parts within a larger system. As I presented in this chapter, there are many authors that treat the subject of informality from the scientific and realistic perspective of the complexity theory. In relation to that, one of the most important things I tried to highlight throughout the chapter is the important shift in the way we look at urban informality today, and "how we acknowledge its profound correlation with the human element" (Di Raimo et al. 2020, p. 2). The human properties, from where the order of the city emerges unpredictably, requires to consider the assemblage or multi-scalar interrelations as crucial to overcome the linearity of spatial analyses (Dovey 2012, p. 387). This refers to new ways of looking at the city in a relationship-based approach rather than as a mass of separated objects.

Then I analyzed the role of the architect in relation to the housing and urbanization issues in the context of an increasing trend of growth in informal settlements. As I tried to make it clear in the chapter, there is a concern for the indifference and even irrelevance of architects and planners toward the vulnerable parts of society because of their staying away from social problems. While there is a clear trend in the shift from the role of the architect as a protagonist in mass housing production that characterized the modern period; to the architect as an enabler helping to integrate its professional expertise with community knowledge in the framework of neoliberal agenda; and to a more discrete role of architect, designer and planner to include adaptive behaviour within a more flexible planning and design practices, as part of an evolutionary paradigm; this trend is not clearly reflected in the education of architects and urban planners. They are not sufficiently exposed to issues of complexity and urban growth, as well as to the topic of informality; especially learned these through practical experiences and in the international environment.

The traces of the natural city can be seen everywhere. The natural city appears in different forms, on different scales: from single buildings, through to neighborhoods and up to the entire city. It appears at different times and stages of a city's history and occurs across cultures. Most importantly, the natural city is not to romanticize the phenomenon of informality by baptizing it under the new label of "natural." It serves to understand emergent urbanism despite being formal/informal as an expression of complex relationships and human properties. In the next chapter, I argue that a more holistic approach is needed to understand the reality of the natural city.

References

Alexander Ch (1965) The city is not a tree. In: Legates R, Stout F (eds) The city reader. Routledge

Alexander C, Ishikawa S, Silverstein M, Jacoson M, Fiksdahl-King I, Angel S (1977) A pattern language—towns, buildings, construction. Oxford University Press, New York

Anklesaria S (2013) Tracing the urban informal—mutual self-help to ExNihilo. In: Abbate A, Lyn F, Kennedy R (eds) The 4th biennial subtropical cities: design interventions for changing climates, paper proceedings

Aquilino MJ (2011) Preface of the book: beyond shelter—architecture and human dignity. Metropolis Books, New York

Ascensao E (2010) The "postcolonial slum"—informal settlements as a "building event" in Lisbon, Portugal. Working paper presented at The Urban Salon. http://kcl.academia.edu/EduardoAscens%C3%A3o

Arida A (2002) Quantum city. Routledge

Ballegooijen JV, Rocco R (2013) The ideologies of informality—informal urbanization in the architectural and planning discourses. Third World Quart 34(10)

Batty M, Longley P (1994) Fractal cities—a geometry of form and function. Academic Press, London, San Diego, New York, Boston, Sydney, Tokyo, Toronto

De Geyter X Architects (2002) After-Sprawl, research for the contemporary city. NAi publishers, Rotterdam; deSingel International Art Centre, Antwerp

Di Raimo A (2020) What does informality have to say to architecture?—decolonizing the enquiry and the enquirer. In: Di Raimo A, Lehmann S, Melis A (eds) Introduction of Part 1 in the book Informality through sustainability—urban informality now. Routledge, London and New York

Di Raimo A, Lehmann S, Melis A (eds) (2020) Introduction in the book Informality through sustainability—urban informality now. Routledge, London and New York

Dovey K (2012) Informal urbanism and complex adaptive assemblage. Int Develop Plan Rev

Jones P (2019) The case for inclusion of international planning studios in contemporary urban planning pedagogy, sustainability

Jones P (2020) The role of adaption in changing the micro-morphology of informal settlements. In: Di Raimo A, Lehmann S, Melis A (eds) Chapter 12 in the book Informality through sustainability—urban informality now. Routledge, London and New York

Kamalipour H, Peimani N (2019) Towards an informal turn in the built environment education—informality and urban design pedagogy, sustainability

Kostov S (2003) The city shaped—urban patterns and meanings through history. Fourth printing, Bulfinch Press AOL Time Warner Book Group, Boston, New York, London

Lexico UK Dictionary. https://www.lexico.com/definition/organic

Malusardi F, Scoppetta C (2008) Rethinking urban environmental-morphology (the case of Rome-East periphery). In: 44th ISOCARP congress. http://www.isocarp.net/Data/case_studies/1152.pdf

Marshall S (2009) Cities, design and evolution. Routledge, London

Merriam-Webster Dictionary. https://www.merriam-webster.com/dictionary/organic#other-words

Mumford L (1961) The city in history—its origins, its transformations, and its prospects, Harcourt Brace Jovanovich, INC, New York

Potsiou CA, Ioannidis C (2006) Informal settlements in Greece—the mystery of missing information and the difficulty of their integration into a legal framework. In: 5th FIG regional conference, Accra, Ghana

Rowe C, Koetter F (1978) Collage city. The MIT Press Cambridge, Massachusetts and London, England

Roy A (2005) Urban informality—toward an epistemology of planning. J Am Plan Assoc, Chicago IL, Spring 71(2):147–158

Roy A, AlSayyad N (Eds) (2004) Urban informality: Transnational perspectives from the Middle East, South Asia and Latin America. Lanham, MD: Lexington Books

Silva P (2016) Tactical urbanism—towards an evolutionary cities' approach? Environ Plan B Plan Design
Silva P (2018) Designing urban rules from emergent patterns: co-evolving paths of informal and formal urban systems—the case of Portugal. IOP Conf Ser Earth Environ Sci 158:012001
Suhartini N, Jones P (2019) Urban governance and informal settlements—lessons from the City of Jayapura, Indonesia. Springer, The Urban Book Series

Chapter 3
Additional Theoretical Lenses to Understand the City

Abstract Understanding the conceptual gap between the natural city and the official planning system requires a fine-tuning of our theoretical lenses to see the broader picture. The refined lenses are the first step to make observable the subtle human interactions and related emerging properties that are the essential underpinnings of real life. For this purpose, in this chapter, I give a short introduction to the theoretical concepts indispensable to viewing these phenomena clearly. These concepts address the issue of interrelationships and their effect on the new emerging realities. First, the idea of interrelationships is broadened by considering them, using quantum logic as the main lens, as actions triggered by material and non-material factors raised beyond local temporal realities. Next, fractal city logic is applied to interrelationships, considered as the recursive relations between elements that create a different kind of order, the structure of which depends on internal dynamics. Finally, using complexity theory, interrelationships are seen as self-organized macroscopic emergent behaviors. However, first, some turning points are presented that overcome the old mechanistic paradigm and envision a different reality from a more holistic perspective.

Keywords Mechanistic · Quantum · Fractal · Complex · Holistic · Inter-relationships · Emergent reality

Understanding the conceptual gap between the natural city (explained in the previous chapter) and the official planning system requires a finetuning of our theoretical lenses to see the broader picture. The refined lenses are the first step to make observable the subtle human interactions and related emerging properties that are the essential underpinnings of real life. For this purpose, in this chapter, I give a short introduction to the theoretical concepts indispensable to viewing these phenomena clearly. These concepts address the issue of interrelationships and their effect on the new emerging realities. First, the idea of interrelationships is broadened by considering them, using quantum logic as the main lens, as actions triggered by material and non-material factors raised beyond local temporal realities. Next, fractal city logic is applied to interrelationships, considered as the recursive relations between elements that create a different kind of order, the structure of which depends on internal dynamics. Finally, using complexity theory, interrelationships are seen as self-organized macroscopic emergent behaviors. However, first, some turning points are presented that overcome

S. Dhamo, *Understanding Emergent Urbanism*, The Urban Book Series,
https://doi.org/10.1007/978-3-030-82731-1_3

the old mechanistic paradigm and envision a different reality from a more holistic perspective.

For decades, scientific evidence has been showing the need to move beyond the mechanistic paradigm. This chapter briefly describes the scientifically supported conceptual leap from the mechanistic view of reality as a perfect machine to the holistic quantum view and its related fractal and complexity theories. Previous work in this area, as detailed below, has informed and inspired the shift toward analyzing and observing cities from a broader theoretical perspective. When applied to the urban analyses field, as shown in Chap. 6, the scientific concepts discussed in this chapter offer an extraordinary transformative conceptual force to view reality differently.

3.1 From a Broken Mirror to a More Holistic View

The rules and apparently precise regulations that govern our cities belong to a partialized worldview subject to a series of limitations, edges, and borders, and with one clearly limited position in time and space. This simplified approach to reality contradicts the natural development of life, contributing more to the fragmentation of our world than to their unification. Educated in this way, in vain we struggle to assemble a complete picture of reality. As Senge (1994 p. 3) said, it is like "trying to reassemble the fragments of a broken mirror to see a true reflection." The task becomes so difficult that giving up without achieving clarity is inevitable. Consequently, demobilization of our skills to capture real-world complexity is the price we pay for these gross simplifications. The same occurs with our understanding of the city. As Campbell and Cowan (2002 p. 21) argued, cities are "victims of specialisms"; good urbanism is about "seeing the whole picture and rediscovering the art of building cities … recognizing cites as complex … environments" and "managing their interaction" (2002, p. 7). While this may sound obvious nowadays, there remains a need to deeply understand the notions of holism and complexity and to apply this knowledge to transform our cities. This leads to another important issue: the crosscutting need to include awareness of the complex and holistic nature of reality in the education of architects and urban planners. In Chap. 2, I referred to many other authors arguing for this need.

The inertia of schismatic and mechanist thinking remains embedded in mainstream ways of thinking. This thinking is grounded in the scientific revolution, which culminated in the seventeenth century with the Cartesian and Newtonian worldviews: one of the major shifts that prepared the world for unprecedented industrialization. Cartesian "scientism" and the "firm belief in the certainty of scientific knowledge" generated an innovative impetus in all fields, but also created some negative long-term secondary effects (Arida 2002, p. 33). As Capra (1982) observed more than three decades ago, we try to understand with outdated mechanistic concepts a reality that does not develop according to those rules. A profound change in mindset, perception, and values was needed.

The myth of scientific precision, based on the Newtonian paradigm of absolute time and space in a universe working as a machine, was fundamentally altered by Einstein's special and general relative theory at the beginning of the past century, which states that space and time are relative and interconnected in a space–time interchangeable continuum. The certainty of scientific knowledge was further eroded during the 1930s when, based on Bohr's complementarity and Heisenberg's uncertainty principles of the quantum world, science accepted its own imprecision (discussed further below). These represent gigantic shifts in the way we think about the world: as a machine with separate objects and subjects, or as a continuum where energy and matter are unified; as a determinist and absolute entity, or as a probabilistic and statistically based one; as superficially chaotic, discontinuous, and irregular, or as a complex intercreative emergence from a multitude of interactions and information exchanges.

As we already know, the city and its parts grow and stratify naturally, not necessarily fitting with the idealistic order inspired by Euclidian geometry and Newtonian determinist mechanics. The dream to precisely plan every centimeter of a city over a long-term horizon is not only impossible, but it is also unnecessary. We have finally realized that this reality of the city is not a fault or something to fight against. Rather, we need to understand the specific urban conditions and related emergent qualities from a broader theoretical perspective. Thus, specific realities are not only products of specific cultures; they depend on how we observe and highlight their properties based on the scientific knowledge and worldviews we deploy, which themselves are subject to continuous change.

Recently, the increasing insistence on the science of complexity as a pivotal conceptual basis has been rapidly reforming the classic mechanistic worldview. For Batty (2013, p. 149), the "new science of cities" will be based on a number of approaches that "adopt and adapt" different methods and tools, which can only happen in the presence of a "consistent philosophy one might describe as complexity." In a later work, Batty (2018) argued that as cities become more complex, skepticism is increasing regarding the ability to predict based on the old certainties, rooted in scientific determinism, that once dominated city planning. These certainties were based on the wrong assumption that city systems were closed off from their external environments. In reality, the complex systems according to which cities develop are inherently unpredictable. Complexity theory deals with systems with emergent structures (raises). This emergence, of "much coming from little," is a ubiquitous phenomenon and an important factor in understanding complex systems (Holland 1998, p. 2). Speaking about complex systems, Mitchell (2009, pp. ix–xii) observed how the reductionist explanation of fundamental physics—which can be traced to the 1600 s and Descartes and Newton—was dismissed only by twentieth-century science. Ensuing antireductionist ideas, typified by the notion that "the whole is more than the sum of its parts," became significant, while new sciences, such as chaos theory and network theory, overcame reductionism.

While this shift in the scientific worldview has been widely embraced, there remains a need in the fields of architecture and urban planning to adopt the quantum paradigm and new science of complexity to inform our observation of the "new" reality of cities as systems with emergent structures. While this reality is not actually new, it becomes tangible only when we observe it and become aware of it. The underlying properties of natural cities already exist; we just need to find the right angle from which to observe them. This new reality is rooted in the firm belief that how we conceptualize the world depends directly on our operationalization of scientific knowledge. The logic of quantum theory, as well as fractal and complexity theory, offer a theoretical framework for observing cities from a different angle, allowing access to the nuances and interactions already occurring in real life.

3.2 Understanding the City from the Quantum Paradigm's Holistic Reality

3.2.1 Emergent Reality and the Wave–Particle Duality

In the book *Quantum City*, Arida (2002) summarized the main concepts borrowed from the field of quantum physics that have contributed to the shift in our understanding of reality. One difficulty posed by these concepts is that they are not intuitive. Thus, to avoid becoming lost in complex technical explanations, in this section I refer mainly to Arida's interpretations of the quantum principles needed for urban observation and analysis; that is, wave–particle duality, Bohr's complementary principle, the Copenhagen interpretation, Heisenberg's uncertainty principle, and quantum field theory. As we will see, the quantum worldview is used in this book to understand reality and its emergent qualities, and to interpret the fundamental phenomena that exist behind the scenes in a quantum world but that are simultaneously present in their potential emergencies. This reality has a profoundly wave-like nature that, while not able to be directly seen, becomes tangible once observed.

The strange behavior or dual nature of reality, both as a particle and wave—that is, wave–particle duality—is the first aspect of quantum theory that challenges our imagination and informs a series of ground-breaking interpretations. In the quantum paradigm, the components of a system "stop being either particle or wave": they can be both (Arida 2002, p. 143). As Bohr postulated in his principle of complementarity, quantum particles exist at once in all their possible states (p. 53), with the "particle-like and the wave-like" aspects of particles "complement[ing] each other" (p. 51). In this way, quantum theory brought back the whole and offered a more complete picture of reality. Further, as seen in the site-specific analyses in Chap. 6, the wave aspect, which is "inseparable from the particle aspect of matter," has a physical influence on reality. A full understanding of an object is not possible if we look at only one of its aspects (Arida 2002, p. 52).

In the 1930s, Heisenberg, the other leading personage in the field of quantum theory alongside Bohr, argued that "it is impossible to measure both complementary aspects [of matter] precisely" in terms of certainty to exist in a definite place and occur at a definite time. Heisenberg tried to regulate using mathematical formulas Bohr's principle of complementarity through his uncertainty principle, which prefigures a "statistically based reality" (Arida 2002, pp. 55–56) showing "tendencies to exist" and "tendencies to occur" (Capra 1982 in Arida 2002, p. 55). Thus, the reality is "interactive" rather than "absolute" and depends "on how an observer chooses to observe it" (Arida 2002, p. 56). This represents a shift from a static reality to an indeterminate, emergent, and nuanced reality, where more is visible than would initially appear to be the case.

Bohr's complementarity principle and Heisenberg's uncertainty principle were the basis for the Copenhagen interpretation, first presented by Bohr in 1920. Among other things, this was an effort to prefigure the world of atoms under quantum physics and establish the importance of indeterminacy in wave–particle duality. Importantly, quantum theory challenges our idea of time and space: the "motion is not continuous," as in classic mechanics, but "proceeds in minute jumps," "transcending time and space." The electron jumps "to all possible states simultaneously checking its possibilities … before 'choosing' a particular state to rest in." Between the different locations, there is an "infinity of possibilities" to which a particle can jump. For the same reason, particles permeate all space and time, gaining "holistic" non-local temporal qualities. This is represented by the wave function, which reflects the "mathematical probability" of the presence of the particle at different points in space (i.e., the Copenhagen interpretation). Under this logic, the reality is infinite possibilities that can be calculated mathematically as probability distribution waves. This demonstrates, albeit unintuitively, the indeterminacy of the quantum world, which clashes with Newtonian deterministic physics (Arida 2002, pp. 53–54). Critically, the Copenhagen interpretation states that "the real world is a mental construct shared by its different observers; the wave function collapses when and how it is looked at; the world exists only when it is observed" (p. 57). As I continuously argue in this book, the new emergent realities and their properties remain on the dark side unless we observe them. This notion is central to how we observe what is beyond the characteristics of the natural city and its related human properties.

Quantum field theory offers another important conceptual leap: that interaction makes the field a physical reality (Arida 2002 pp. 61–62). This interaction is due to the non-local temporal properties of the quantum particles transcending time and space (p. 53). To illustrate this interaction, Arida used the duality of the "event" (punctual in ST) and "associated event horizon" (the associated wave) as an interpretation of wave–particle duality, or material and non-material aspects, which can also be easily applied during the analytical process. Thus, because of the associated wave effect of an event that can be more than one, "its territory can spread [emerge] to the limits of its event horizon" (p. 149). The event horizons of one or more dualities can intersect and overlap, producing "interference patterns" (p. 150), which differ from the original waves (p. 151). Considering the non-local temporal properties of an event, its effect can also be felt outside its event horizon. As seen in the next chapter, the human user and their memory transmit these non-local effects (p. 153).

By applying these quantum concepts to the urban field, complex and interesting phenomena emerge. In analysis, this encourages the identification of reasons that extend beyond a territory's local temporal frame. A territory is no longer a physically limited entity: it can exceed its spatial and temporal limits and exist beyond its local time and space. Thus, the territory or city is considered a "malleable energy field" that inclines the system to change (Arida 2002, p. 150). Because a new quality emerges when two quantum systems meet, Zohar and Marshall (1994, p. 63) called this the creativity of the relationship that exists in quantum reality. Emergence, which is a critical issue in urban processes, cannot be fully understood without first considering these concepts.

3.2.2 Multiple In-Between Gradients

The complementary principle and the indeterminacy of the dual aspects of matter upset existing conceptualizations of determinacy and precision, which traditionally corresponded to a rigid and exclusivist approach to reality. A fine example of the colorful palette fostering the transition from an exclusivist to an inclusive approach—or in Arida's (2002, p. 142) words, from a dualistic "either/or" to a tolerant dual "both/and logic"—is offered by the reality of our cities. As such, the infinity of transitional nuanced categories should not be rejected simply because they do not comply with the rigidity of the urban planning laws or our mental customs.

Wave–particle logic—in which particles are defined in space and time, while waves are a dynamic non-local temporal quality transcending space and time—can be extended to other dualistic categories (Arida 2002, p. 143). As discussed in Chap. 2, in relation to the natural city as complementary nuanced variety, the actions undertaken by people as part of the mute negotiation process at the smallest scale, in which they test permissibility to approach one or the other "extreme" statuses of a dualistic category such as public–private or private–private, create in-between categories from which complementary, uncertainty, and indeterminacy arise. Therefore, instead of categorizing something as public or private, there is a wide range of complementary and nuanced variations between these extremes, corresponding to each individual or group action. As such, these dual categories obtain more meaning in specific contexts and gain the quality of a continuum, whereby changes in one influence changes in the other. Thus, they are not separate or dichotomous but have an emergent nature in their complex interrelations.

The tolerant and complementary approach introduced by quantum theory is key for understanding the new emergent qualities of natural cities. From this angle, this book is concerned with the dual category of the natural city and human properties, with the intent of identifying the enzymes inherent in the mutual interchangeability and complex relationships within this continuity. The identified patterns (presented in Chap. 5), as new emergent qualities, are expressions of this duality. This duality is further detailed in Chap. 6, in relation to anthropological principles and the complex pattern generation processes. Using the logic of duality and in-between gradient meanings, we can expand the basis from which to observe and analyze cities by including previously omitted categories, allowing the detection of the infinite potentials to which cities are inclined. Even in cases considered problematic (e.g., informal settlements), this approach motivates alternative solutions based not only on philanthropic moralities but also on an awareness of reality as a mental construct.

Having been exposed to these concepts, we cannot but have remorse about how we do in relation to the inadequacy of current planning and design approaches, in which each element is thought of as a single particle delineated in time and space, becomes clear. Instead of speaking about the interaction of a multitude of factors, planning continues to focus on dichotomies. In its current form, the rigid planning system tries to "correct" reality, and thus life itself, without considering that the problem lies in the outdated paradigms on which the system is based. As I have argued throughout, while current legal and regulatory frameworks take the deterministic approach, uncertainty and indeterminism are ubiquitous in real life. Deterministic laws have been translated, in reality, into spontaneity and informality, fostering a non-compliance culture. This has been illustrated in this book by the administrative attitude toward the informal city. In this regard, a non-deterministic approach, in the quantum sense, may enable self-organization and self-regulation based on understanding and steering processes that seed the germs of human properties and are more permissive to an open-ended result.

While the shift from physical determinants and social structures to material and non-material aspects may sound like a matter of jargon, this is in fact an important conceptual shift. The shift to generalized categories enables the introduction of more categories (or nuances), which are useful for enlarging the conceptual base for observation and analysis of urban reality, as shown in Chap. 7.

3.3 Understanding the City from the Fractal Perspective

3.3.1 Relational Emergent Order

One of the gross simplifications we make in regard to the city concerns its physical form. When discussing shape, irregularity normally implies a lack of uniformity in Euclidean terms. This makes sense within a limited scientific understanding; however, in this section, I argue that, viewed according to fractal geometry, what is normally considered irregular does not exclude the emergence of a different kind of order. The stigma of irregularity in regard to the city form reflects and motivates the desire to impose a centrally controlled artificial geometric order based on excessive simplification. Speaking about natural cities, I already touched upon this logic to challenge the idea of their irregular form and to redefine the concept of the spontaneous, informal and organic city as a different kind of order. In this section, this concept is elaborated by providing details about how irregularity can be differently perceived. Instead of considering the natural city unplanned and irregular, the focus is on how they coordinate at the individual level.

Reframing chaos as a more complex form of order without eradicating layers of real-life constitutes a conceptual step forward in understanding reality. In this regard, the principles of "fractal cities," summarized by Batty and Longley (1994), are crucial to enlarging the concept of irregularity in relation to form and shifting from an understanding of the geometry of the city in pure Euclidean terms to seeing the deeper emergent order beneath. For Batty and Longley, fractal geometry, which also signifies the "geometry of chaos" as developed by Benoit Mandelbrot at the beginning of the 1980s, plays an important role in city analysis and simulation. A deeper understanding of form in relation to "process, scale and shape, … statics and dynamics" (p. 10) provides an additional theoretical lens through which to overcome simplifications. This also supports a shift away from the "doctrine of visual … order" (p. 8) and "physicalism" (p. 1) that considers the city as "architecture-writ-large" toward an understanding of cities as systems of organized complexities (p. 1), which is another thesis supported by and elaborated in this book. As Batty and Longley argued, many preexisting theories and urban models are based on fractal logic. In fact, the use of these principles is important for understanding cities of any time and any culture (p. 33).

3.3.2 Principle of Scale, Repetitions, and Self-similarity

In the natural city, the geometric principles of the form "vary with respect to scale." What appears to be "ordered in terms of pure geometry" at one scale might not appear so at the smaller scale. An example of this is the geometric layout of apartment blocks, which then gives way to individual arrangements on a smaller scale. Zooming in and out is thus applicable in fractal geometry to explore the different orders hidden at each level of the city hierarchy (Batty and Longley 1994, p. 33). In this sense, the irregularity seen at a glance in many natural cities is a superficial manifestation of a deeper order.

The fractal generative process, which is infinitely scalable, is defined in terms of the "initiator" as a starting geometric object; the "generator" as a modality applied to the initiator in a repeated way through scales; and the "cascade," which controls the application process. By "applying the generator to the initiator," the resulting geometric object is "composed of several initiators at the next level of hierarchy" (Mandelbrot 1983 in Batty and Longley 1994, pp. 61–62). This recursive process creates the property of "self-similarity" (Batty and Longley 1994, p. 63), which emerges from the repetition of many common actions applied on elements that consequently follow specific behaviors across spatial scales (or at different levels of the hierarchy). Thus, "if similarity is strong in a geometric sense, then it is referred to as self-similarity or in its weaker form as self-affinity" (p. 47). Most natural objects manifest "self-affinity." This property is created by stretching or distorting the object in one direction during the process of applying the generator to the initiator (p. 63). The secret of the rediscovered order, in what initially appears chaotic but in fact contains a hidden and complex legibility, remains in these two properties, which are inherent in the shape and form of cities. For example, "objects which show the same kind of irregularity at many scales," called "fractals" by Mandelbrot (1983, 1990), are irregular in Euclidean terms because they are not smooth. However, ordered patterns are in fact displayed within this irregularity (Batty and Longley 1994, p. 59).

For example, in the case of Tirana, after the organic growth phase, a geometric radio-centric scheme was laid over the city. Most of the developments at the urban scale (e.g., the infrastructure, and important buildings) respected that scheme in parallel with the previous organic one. Here, the logic of scalability helps to uncover the hidden emergent order at different levels, elucidating that aspects not making sense at the finer scale (e.g., interventions relating to the broader infrastructure system of the city) are part of a larger scale order.

3.3.3 Form as a System Structure

From the discussion thus far, we can see that the self-similarity and self-affinity of fractals emerge from a recursive process and that these properties generate a hierarchical structure. The main idea of fractal cities is to identify "systems in which elements are repeated in a similar fashion from scale to scale" (Batty and Longley 1994, p. 47). Thus, for a fractal understanding of the form of cities, first, we need to know the elements of the system, how these elements are grouped and repeated in subassemblies through scales, how these groups create self-similar or affinitive hierarchical structures, how the entire structure emerges and transforms from relations and forces interacting within it, and the internal dynamics of the form. Related to that, White (in Batty and Longley 1994, p. 43) discussed "external form" and "internal form," with the former representing what is visible and subject to direct observation, while the latter corresponds to the structure. Both define the "spatial form." In fact, investigating the urban form in terms of its "structure" and "behavior" reveals a different kind of order—the order beneath, emerged by invisible relations. Thus, considering the city as a diagram of forces with its statics (structure) and dynamics (behavior) goes beyond a merely physical approach and enables the investigation of the city through the lenses of organized complexity. These lenses are not geometrically abstract; they fit everyday processes and explain their human nature, helping to give meaning to the city as traditionally known.

These principles are important in how we see reality: in terms of simple Euclidean geometries that cannot address issues related to the generative process or the process of growth, or in terms of more "abstract geometrical relations, hierarchies and networks" that may produce a more holistic meaning (Batty and Longley 1994, p. 55). This geometry enables one to read relational order and structure through scales, by offering some guidance for viewing the organized complexity that emerges from relationships between traces of growth in a palimpsest, or to see the record of events and their related horizons in a series of overlays. As already mentioned in the previous section, these horizons may emerge even in the presence of "absence," in the sense of the non-local temporal properties of an event. This conception extends beyond simplistic and superficial interpretations of the form of cities. In Chap. 6, which is dedicated to pattern analysis, attention is directed to understanding form as a "system with structure." This is considered an underlying form that helps to explain what we see in the external urban shape (Batty and Longley 1994, p. 43).

3.4 Understanding the City from the Complexity Perspective

3.4.1 Self-organization and Emergent Realities

The top-down approach by which we currently design and manage our cities reflects another gross simplification in regard to the bottom-up self-organized properties inherent in the city. According to Portugali (2000, p. vii), "self-organization is the central property, of … complex systems," something I emphasized in Chap. 2 with regard to the natural city. Indeed, self-organization is the starting point and main "reservoir" in which to search for the human properties embodied in the natural city and its emergence. In addition, self-organization provides the conceptual basis for understanding the complexity of the always-changing urban environment. Thus, this property turns out to be a crucial concept for the topic we are discussing and emergent urbanism.

According to Mitchell (2009, p. 4), a complex behavior is "how large numbers of relatively simply entities organize themselves, without the benefit of any central controller, into a collective whole that creates patterns, uses information, and in some cases, evolves and learns." For example, similar complex behaviors can emerge from the relationships within self-organized groups during the settlement process. In this case, the multitude of individuals interacting in a process, or involved in a macroscopic self-organized behavior called emergent, is not centrally controlled but triggered by a common or competing motive, based on simple rules and exchanging information. Interestingly, as covered by Chap. 6, intelligence or consciousness emerges from these processes; thus, elements in the system learn and evolve. The challenge remains to understand how, from "underlying simple rules" (p. xii), complex systems produce interrelations, signaling, and re-adjustable behavior over time.

Arida (2002, p. 174) saw self-organization as the capacity of a system to "borrow energy from [its] environment" to become self-regulating, "giv[ing] rise to higher order complexity." Only "open systems" that exchange energy or matter (or information) with their environment are capable of "complexification and differentiation and hence of diversification" (p. 174). Therefore, the concept of a system exchanging energy with the environment becomes crucial here. If we accept the non-local temporal properties of an event (event–event horizon) as presented in the previous section of this chapter, there are infinite ways in which this exchange may be triggered and materialized (emerge) in a new territoriality. In terms of the fractal approach, the environment owns scalability properties created by repeatable actions or relationships through scales. If we apply these two concepts to the ambit of the urban environment, the situation becomes even more interesting and complex. In Chap. 6, I address this issue further both in my argument that urban patterns are complex phenomena and by applying quantum and fractal approaches to urban analysis.

According to Holland (2014, p. 26), each complex system exhibits a "property called emergence," summarized in the phrase that "the action of the whole is more than the sum of the actions of the parts." In this sense, the issue is to identify how systems "exhibit emergence" and how "to attain some ability to 'steer' the complex system" (p. 27). For Holland, another property of complexity that is closely related to emergence is "hierarchical organization" (p. 29), where each level is regulated by its own set of rules. This kind of system is pervaded by top-down and bottom-up actions (pp. 29–30) and characterized by behaviors such as "self-organization into patterns," "chaotic behavior," and "adaptive interaction" (p. 31).

Investigating how elements of a system interact and adapt, and how relationships emerge while they self-organize into patterns, is crucial for the analyses presented in this book. This way we obtain a more holistic understanding of the emergent realities where relationships are involved, or in the words of Holland (2014, p. 28), where "the aggregate exhibits properties not attained by summation." First, let us briefly reflect on the meaning of patterns emerged as the underlying structure of relations.

3.4.2 *Urban Patterns as a Complex Behavior*

According to the Merriam-Webster dictionary, a pattern is a "natural or chance configuration." This meaning inherently bears the quality of emerging. Second, it is defined as a consistent "sample of traits, acts, tendencies, or other observable characteristics." This reflects the idea of order emerged by similarly repeatable actions based on a specific logic. A third definition is "a discernible coherent system based on the intended interrelationship of component parts." This implies that patterns emerge from the interaction of a multitude of elements. Finally, patterns are defined as having a "frequent or widespread incidence." This implies that certain actions are repeated more frequently than others, and that statistical correlations can be revealed in the way the system behaves. This definition anticipates the idea presented in Chap. 7 that urban patterns own computational and algorithmic information.

From this perspective, and from what we presented in the previous section, we can conceptualize the urban patterns as the capacity to self-organize by borrowing energy from the environment and ideally evolving toward self-regulation, giving rise to a higher order complexity. As such, the urban patterns own site-specific essential qualities emerging from the underlying structures of interrelationships, or repetitive actions triggered by human motives, which transform in social or group norms through information exchange and communication to solve everyday problems. These interrelationships may give rise to a specific pattern. Referring to what we presented in the previous section about quantum theory, patterns are local temporal manifestations existing within a larger continuum or correlations in space and time. Therefore, patterns are not static but are transformed through space and time and by specific social conditions. Patterns can also be considered morphological expressions (or, to use the term appearing later in the book, "horizons") composed of clear formal sequences or an emergent order made of elements and relationships in a system. This is significant for the way specific neighborhoods are generated.

The logic of pattern is also used by Alexander et al. (1977) in *A Pattern Language*. For these authors, the logic of patterns is distilled from practice that values life experience. Each pattern is a timeless entity (pp. IX–XXXV) representing the "best guess" about the arrangements of the physical environment to solve the problem (p. XV). According to Alexander et al. (1977), each of the 253 patterns they propose is connected to other patterns, creating "an infinite variety of combinations" (p. XI). These connections have the "structure of a network" (p. XVIII), comprising sequences from which we can "make a language" (p. XIX). For these authors (1977), "each pattern describes a problem" while simultaneously describing "the core of the solution to that problem" (p. X). Interestingly, while the problem "describes the empirical background of the pattern," the solution describes the "field of physical and social relationships which are required to solve the stated problem" (p. XI) to allow for adaptations to local conditions. This is similar to how I view patterns. While Alexander et al. speak about a problem and solution described by patterns, I speak about motive as the triggering factors and modality as social norms that give rise to a pattern (Chap. 6). As such, patterns rather than proposing schemes for design, are a potential informative, generative, and processual logic for design.

As seen in Chaps. 5 and 6, the emergent realities of the analyzed urban patterns and their essential qualities generated by interrelationships are an illustration of such complex reality. Looking at patterns as a structure of human relations or interactive processes that generate that form, we go behind the physicality of form that treats the city as an empty box or part of an inanimate world. As such we attribute human properties and meaning to urban patterns and overcome the simplifications. It is evident that self-organization is the observable phenomenon from which we can start identifying the self-generated new emergent realities of urban patterns, which have interrelationships at the base.

3.5 Conclusions—Emergence as a Key Concept

In the first section of this chapter, we discussed two ways of looking at reality: as resulting from the mechanistic sum of its constituent parts and deprived of meaning, or as emerging from the overlap of the wave function, where relationships triggered by specific motives play a principal role. For this, some key points presented from quantum theory started to offer a clearer picture about what we can consider holistic reality emerged from internal relationships of the wave aspect. Because of that aspect, the territory is not a physically limited entity, it can spread beyond its local temporal reality. In this sense, the territory can be considered a malleable energy field in continuous transformation. Thus, the quantum notion of wave–particle duality and its associated ideas of interactivity, inclusive complementarity, uncertainty, and indeterminacy helped to reconceptualize the way we look at reality.

Fractal perspective helped to further refine our visualization of structure and dynamic of the "new" order emerging from relationships. It provides a means of seeing that the degree of order is deeper and emerges from an invisible structure of relationships and hierarchies, as well as recursive irregularities that repeat themselves geometrically across scales. This creates a new logic of regularity based on the properties of self-similarity and self-affinity. This way the fractal cities approach aims to overcome the concept of disordered cities accepting many types of regularity, which emerge from relationships.

The lenses of complexity theory help to discover how the self-organized systems exhibit emergence and the set of rules that combine bottom-up and top-down cyclical processes that repeat and generate certain recognizable patterns. Emergence as a key concept in complexity theory cannot be understood unless we first consider the concepts presented in this chapter from quantum and fractal theory through which we can give structure and dynamics to this reality. In the case of a self-organized settling process, the dynamics of individuals and social groups are materialized in concrete actions and processes happening in the real life.

Finally, I tried to conceptualize urban patterns as entities emerging from a complex behavior. In relation to that, urban patterns could be observed and analyzed as wholistic entities emerging from quantum correlation effect; as containing an emergent order generated by similarly repeatable actions within a system structure; and as relationships emerging in a bottom-up self-organized collective whole that creates structure ad patterns. This conceptualization is the base for the analyses in Chap. 6, where urban patterns and their morphological expressions are observed and analyzed as emerging from an underlying structure of interrelationships.

This chapter is an important step forward in relation to new ways of looking at the city. The basic theoretical concepts presented in the chapter, help to move away from simplistic interpretations and to understand emergence as a key property of the complex systems. The natural city and its emergent qualities can only be detected if the theoretical lenses and analytical methodologies applied to observation enable this. The concepts presented thus far extend our observation field, allowing for a more holistic understanding of reality. Only in this way will the often-jargonistic term "holistic" carry real value.

References

Alexander C, Ishikawa S, Silverstein M, Jacoson M, Fiksdahl-King I, Angel S (1977) A pattern language: towns, buildings, construction. Oxford University Press, New York

Arida A (2002) Quantum city. Routledge

Batty M (2013) The new science of cities. The MIT Press, Cambridge, Massachusetts, London

Batty M (2018) Inventing future cities. The MIT Press, Cambridge, Massachusetts, London, England

Batty M, Longley P (1994) Fractal cities: a geometry of form and function. Academic Press, London, San Diego, New York, Boston, Sydney, Tokyo, Toronto
Campbell K, Cowan R (2002) Re:urbanism. Urban Exchange, London
Capra F (1982) The turning point: Science, society and the rising culture. Wildwood House
Holland John H (1998) Emergence from chaos to order. Perseus Books, Cambridge, Massachusetts
Holland John H (2014) Complexity: a very short introduction. Oxford University Press
Merriam Webster online vocabulary. https://www.merriam-webster.com/dictionary/pattern
Mitchell M (2009) Complexity: a guided tour. Oxford University Press
Portugali J (2000) Self-organization and the city. Springer-Verlag, Berlin Heidelberg GmbH
Senge Peter M (1994) The fifth discipline: the art & practice of the learning organization. Currency Doubleday, New York, London, Toronto, Sydney, Auckland
Zohar D, Marshall I (1994) The quantum society: mind, physics, and a new social vision. Quill William Morrow, New York

Chapter 4
Historical Notes on Tirana

Abstract In this chapter, I give more details on the history of the urban processes in Tirana in relation to political influences, design, and planning types, and how they materialized in public and private spaces. We begin by considering Tirana's foundation as an organic settlement at the beginning of the 17th century under the strong influence of the Ottoman Empire, before then reflecting on the attempts of King Zog and the Fascists to Europeanize Tirana. From this focus on the aggressive interventions by the totalitarian regime to design the city through demolition and geometric reshaping, we shift to examine the explosion of the urban form and return to spontaneity witnessed after the collapse of the regime. This chapter analyzes the fluctuations in Tirana's urban fabric and attributes the multifaceted reality to the transformative historical, social, and political events through which the city has lived. Three main urban patterns are distilled in this chapter as generations and emergences formed under specific historic and social conditions.

Keywords Tirana · Organic city · Ottoman · Europeanization · Dictatorship · Self-organized

In Chap. 1, I presented a general overview of Tirana as a case study to observe and analyze notions of self-organization. As mentioned, moving between spontaneity and authoritarianism produced an antagonistic energy that played an important role in shaping Tirana. In this chapter, I give more details on the history of the urban processes in relation to political influences, design, and planning types and how they materialized in public and private spaces. We begin by considering Tirana's foundation as an organic settlement at the beginning of the seventeenth century under the strong influence of the Ottoman Empire, before then reflecting on the attempts of King Zog and the Fascists to Europeanize Tirana. From this focus on the aggressive interventions by the totalitarian regime to design the city through demolition and geometric reshaping, we shift to examine the explosion of the urban form and return to spontaneity witnessed after the collapse of the regime.

Tirana is a city that shows itself only when the "curtains" separating one scene-event from another are drawn back: the fragments of the city representing various characters are hidden and wrapped inside each other, starting from the outer space along the main streets, to the most hidden and secret heart of the neighborhoods (the

S. Dhamo, *Understanding Emergent Urbanism*, The Urban Book Series,
https://doi.org/10.1007/978-3-030-82731-1_4

mëhalla[1]). The remnants or traces from previous periods interlock in an intricate and complex mass, resulting sometimes in accidental mechanically created offcuts and sometimes in a more consolidated urban fabric. Whatever the case, the created spaces are always full of surprises, creating a discontinuous continuum. Therefore, Tirana can be characterized as a natural city not easily shaped through plans. However, some urban patterns are clearly shaped and legible in Tirana. In this sense, this chapter presents the historical and social context/continuum generating the system of relationships underling the urban patterns, which are further defined and analyzed in Chaps. 5 and 6.

4.1 Transformations and Events

See Fig. 4.1.

4.1.1 The Emergence of the Organic City (1614)

Tirana started as an organic settlement at the beginning of the seventeenth century (Frashëri 2004, p. 17), at which time this part of the Balkan territory was dominated by the Ottoman Empire. Notes from Hahn, the Austrian Consul in Albania, in 1854, give the size of Tirana as 2,000 houses, of which 100 were Orthodox Christian, 6 were Catholic, and the rest were Muslim (Hahn 1854, 2007, p. 124). It remained a small and insignificant center until it was declared the capital of Albania in 1920. Historians and other scholars agree that the first urban nucleus in the Tirana valley emerged close to the Lana River, at the intersection of the commercial routes emanating from the more important regional centers of the time (Fig. 4.1). According to Aliaj et al. (2003, pp. 14–15) and Frashëri (2004, pp. 161, 67, 71–72, 169–173), the first nucleus was founded around a complex containing religious and service institutions, including the Old Mosque of *Sulejman Pasha* (demolished in 1944), the bakery, the *hamam* (demolished in 1938), the inn (demolished in 1958), and the Old Bazaar (demolished in 1958). Other nuclei emerged in relative proximity to this initial site, within a radius of approximately 1 km (Fig. 4.2).

The location of these nuclei in the Tirana valley and their gradual growth reflected and embodied the qualities of their geographic container; that is, the valley's central position in relation to the surrounding mountain and hill ranges—which provided protection and a natural enclosure—and its clear spatial orientation and identification. In addition, proximity to multiple rivers offered plentiful access to water. The site

[1] *Mëhalla* is the Albanian word for neighborhood; the same word *mahalle* is used in Turkish borrowed from Arabic language, from where the original word comes. *Mëhalla* infiltrated in the Albanian language after the Ottoman occupation. The use of this word in the context of modern Albanian carries the meaning of a traditional neighborhood, often associated with the Ottoman background.

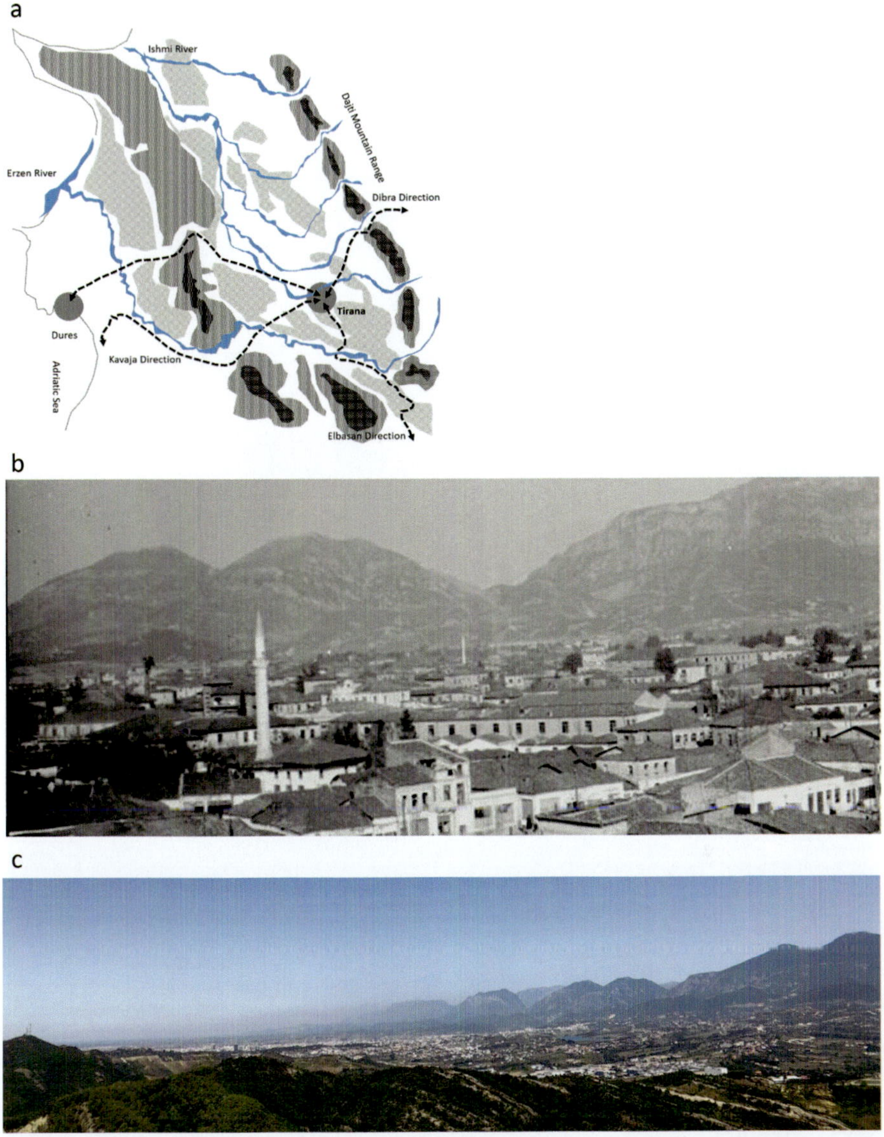

Fig. 4.1 **a** The geographic container (the "Eden") showing the emergence of the organic nucleus of Tirana at the intersection of the regional trajectories from Durres, Dibra, Elbasan, and Kavaja. The two views of the Tirana valley below show (**b**) the city's historical nucleus at the beginning of the 1920s from close up (National Technical Archive [NTA]); **c** a distant view of Tirana's further development as at 2020

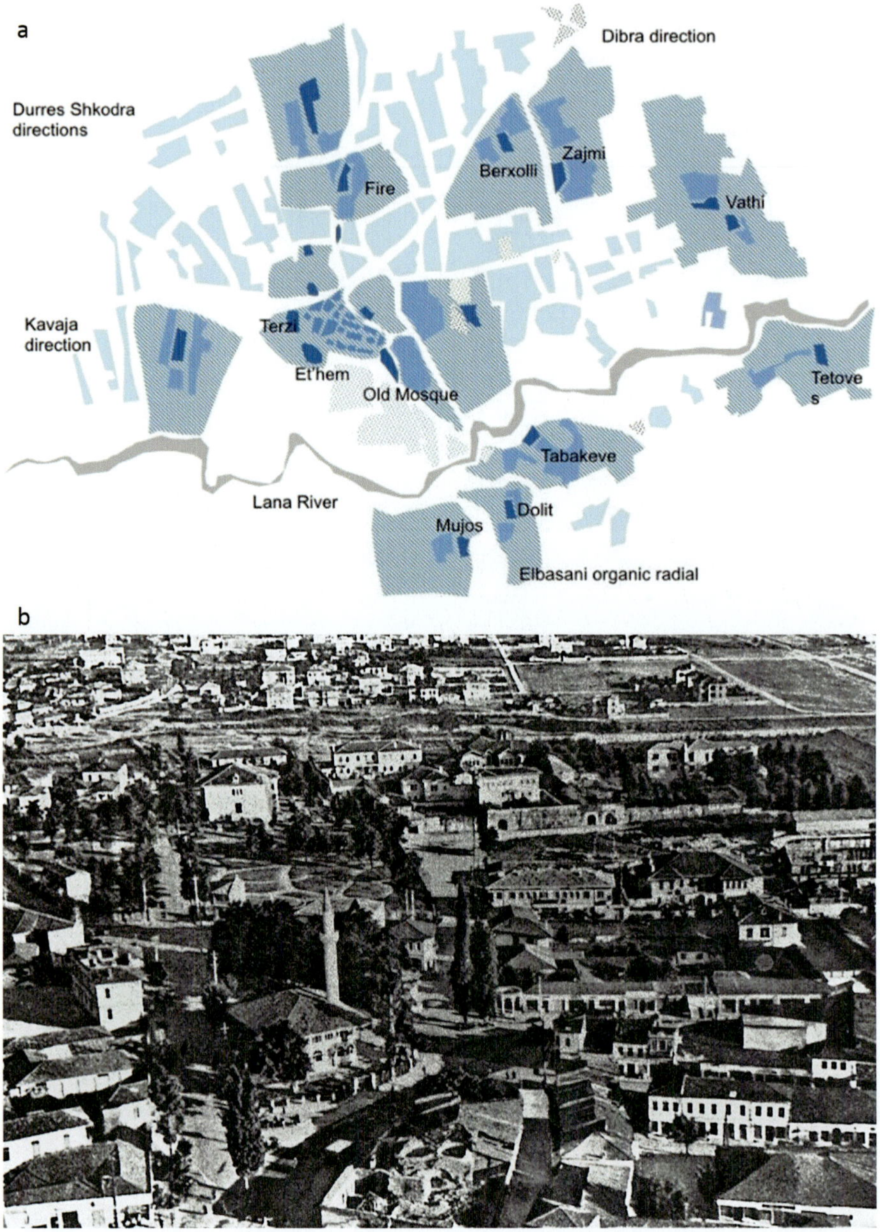

Fig. 4.2 **a** Schematic of the first nuclei of *kulliye(s)* and the gravitation of the masses of houses outlining the first organic layer around the old Bazaar. After the establishment of the first nucleus of the Old Mosque, other nuclei emerged, each of which was named after its donor's family. The second nucleus emerged around the *Fire Mosque* at the beginning of the eighteenth century. This was followed by the emergence of the nuclei of *Zajmi, Haxhi Et'hem,* and the others (Aliaj et al. 2003, pp. 19–20). **b** The first nucleus of the Old Mosque at the beginning of the past century (NTA)

was both the center of the territory and of the entire universe; it was a sacred place that carried all the potential charge for the future city; it was Eden itself (Fig. 4.1). As described by Frashëri (2004, pp. 73, 80), the inception of the first nuclei and their further growth were influenced by the excellent qualities of the geographic container. He refers to Ottoman chronicler Çelebi ([1660–1661] 2004, p. 75), who describes that the vineyards and gardens are innumerable in Tirana; and latter to Hahn ([1850] 2004, p. 180) who writes that the city lies in the middle of fields, gardens, and orchards that are very well maintained.

This process, while natural, was not entirely undirected. During the seventeenth century (and until 1912), the Tirana valley was dominated by the Ottoman Empire. Thus, the abovementioned organic process occurred alongside the *imaret* system, adopted by the Ottomans in the fourteenth century. This system was rooted in the social and cultural principles around which Ottoman life was organized and used the *waqf* (charitable institution) as a form of "Muslim propaganda to [control] a predominantly Christian population ... through devote and moral religious conduct and works of welfare" (Ingersoll and Kostov 2013, p. 439). Essentially, the Ottomans promoted urban life through this system of *imarets*, which included religious complexes (a mosque, or sometimes a *tekke* for Sufi Dervishes), a *turbe* (tomb of the donor), religious schools, baths, sometimes a hospital, markets, and an *imaret* (public soup kitchen), from which these complexes derived their name. The urban structure of Tirana was underpinned by this way of life.

Most importantly, from a city-formation perspective, the *imaret*—later renamed *kulliye* in the twentieth century to emphasize its public function and dissociate it from the soup kitchen (Ingersoll and Kostov 2013, p. 439)—was a complex of service, administrative, and economic activities. The *kulliye*, meaning "the whole" in Arabic, clustered around the mosque, which becomes the center of a neighborhood identified by family ties, profession, or origin (Kostov 1995, p. 457). In Tirana, these first public buildings, once distributed across the Tirana valley, served as pre-urban nuclei and centers of clearly defined *mëhalla* (neighborhoods), around which were organically assembled residential areas (Fig. 4.2) based on "sociological, property and family connections" (Aliaj et al. 2003, p. 18). These *mëhalla* comprised dwellings, barns, courtyards, wells, green groves, lane squares, and graveyards. The preexisting natural or artificial territorial elements played an important role in the location of these elements. The city grew around these initial points or "primary elements" (Rossi 1984, p. 82) a single artifact or nucleus.

Given these historical circumstances, Tirana's predominant ideology and culture were dictated by the Ottoman Empire, which sought to expand its territory in the Balkan Peninsula and beyond. The *imaret* system, and other lifestyle principles based in the *Qur'an*, underlay the generation of many parts of the city, by tracing an indelible imprint for future events in Tirana.

Over time, the sociological units, or community clusters, that comprised the *kulliye* system merged into a dense urban core centered on the most important public buildings of the time: the Old Mosque and Old Bazaar. This process was somewhat similar to what Kostov (2003, p. 60) described as "settlement coexistence" or the first phase of "synoecism," an Aristotelian term meaning "living together" (p. 60). According to

Kostov, "Synoecism ... enables people to transcend their tribal/pastoral ways" (p. 60) and is one of the most common founding processes in Europe, Africa and the Near East (p. 62). In Tirana, the synoecistic center absorbed the scattered nuclei and intervening open spaces, while leaving sites such as markets, bazaars, and graveyards for communal activities. The cohesive nucleus was characterized by an extensive organic layout, conditioned by the fragmentation of feudal land ownership. More than simply a system of laws guiding the process, the *imaret* as a *waqf* institution influenced the entire urban structure and way of life.

From a morphological perspective, three main vector forces contributed to the plan of Tirana as a basic originating element where the urban tissue knitted: the centripetal regional fluxes that collided in the center and generated the Old Bazaar, and the secondary centrifugal and centripetal fluxes from and to the Bazaar. This energy created the third set of vector forces, the wrapping strings, which behaved like a magnetic field around the nucleus of the Old Bazaar. (Using the concepts presented in the previous chapter, this phenomenon can be considered as an "event" with an event horizon.) For this reason, the first nuclei of *imarets* settled not only along the main centripetal radials but also along the wrapping strings around the Bazaar. Filaments following the rotating force lines, intersecting with the centripetal and centrifugal fluxes, connected and knitted the previous nuclei into a concentric urban mass (Fig. 4.3). The interaction of these forces, combined with the preexisting physical determinants, influenced the city's modality of growth around its center.

Fig. 4.3 Diagram drawn over a map from 1921 showing the centripetal regional fluxes, secondary centrifugal and centripetal fluxes, and the wrapping lines around the main nucleus of the Old Bazaar

These characteristics of the plan, and its "persistence" (Lavedan in Rossi 1984, p. 51), have played a role in shaping Tirana and remain the basis of the city organism, transmitting signals even into the future. In this sense, we can consider this as an interfering pattern that emerged within a continuous whole.

The most frequently accessed public spaces within residential areas tended to be the mosque, *teke*, or church (for local prayer services), graveyards, markets, and baths. In the case of Tirana, most public gatherings occurred in the organic center created by the Old Bazaar, which housed at least three central mosques. The public buildings of this space were interlocked within the urban maze, creating interdependent systems within the center itself and with the rest of the city.

The generation and modalities of growth formed under this condition are analyzed in Chaps. 5 and 6 under Pattern 1: Historical Organic.

The primary streets led to the Old Bazaar: the site of retail and wholesale trade, with artisans clustered by street. Secondary streets flanked mostly by residential buildings ran through the *mëhalla,* feeding the dead ends of the lower levels. The dead ends penetrated the secret core of the *mëhalla* and could even be closed off by doorways where they intersected main streets. The crossing of two or more streets resulted in open-air rooms of irregular shape.

4.1.2 King Zog and Stile Littorio

King Zog who ruled Albania from 1922 to 1939, began projects for the Europeanization of Tirana during the early 1920s. This was the first effort to restart the city based on completely different principles and represented the first layer of disturbance aimed at moving from spontaneous growth to centrally designed and controlled town planning and administration. This process was initiated by stretching and enlarging the main organic avenues, such as *Rruga e Durrësit* and *Rruga e Kavajës*, and opening the new boulevard of *Zogu i Parë* (Fig. 4.4). Thus, alongside the organic tissue of the city, a designed and substantially different layer was laid down.

The most significant project was the ceremonial complex of the ministry square and the central axis of the new boulevard, first designed by the Italian architect

Fig. 4.4 The "ideal city" viewed from the main square (NTA)

Armando Brasini in 1925. This project was highlighted as the most important element and the symmetric axis of Tirana in both the regulatory plan of 1929 formulated under King Zog and the regulatory plan of 1939–1943 formulated during the Italian Fascist occupation (Figs. 4.5 and 4.6). These plans based their designs on a philosophy and mentality distinct from the Ottoman heritage of Tirana, creating a new base for shaping the city's future developments. For the first time, a zoning structure was established, combined with a ring-radial system aimed at bringing continuity, both in new areas and in existing organic ones. This thus represents an intentional shift away from the principles and philosophy upon which the previous growth of the city had been based. From this point, Tirana could be identified not only by its dead ends and narrow, winding streets, but also by the straight and open axial boulevard.

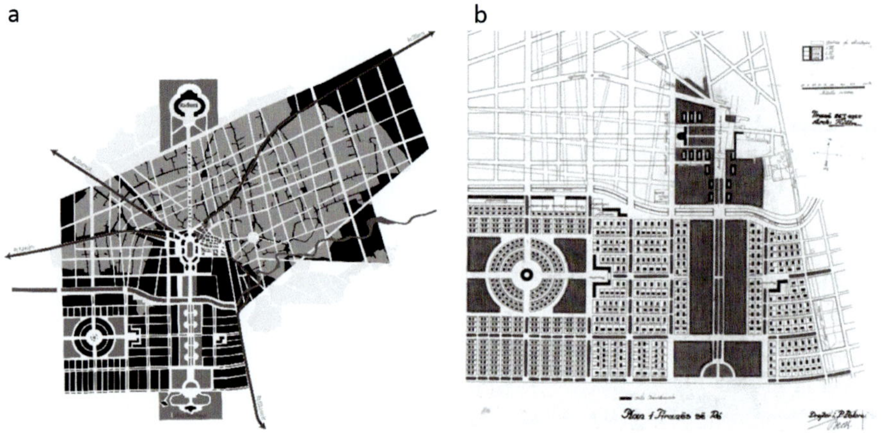

Fig. 4.5 **a** the 1929 plan and the relation with the axis, authored by Frashëri, Di Fausto, and Kohler; reproduced from Dhamo et al. (2016), based on NTA sources. **b** the New Tirana plan of 1928, designed by Kohler (NTA)

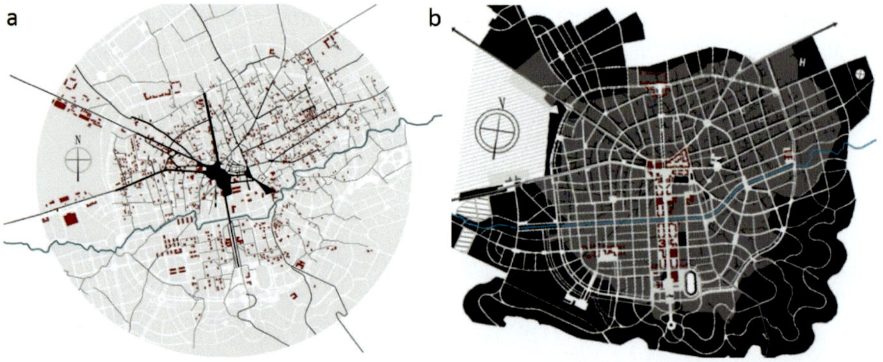

Fig. 4.6 The 1939–1943 existing situation (**a**) and plan (**b**), authored by Bossio; completed by Lambertini and Poggi; reproduced from Dhamo et al. (2016), based on NTA sources

The design concept and meaning of this boulevard, as the main element introduced in the city plan, can be analyzed in relation to topography and its power of representation as a political diagram. Importantly, beyond merely being an abstract geometric scheme and ceremonial or political diagram, the boulevard was well adapted to its situational context. This axis represented the meeting place between the city spontaneously emerged and the designed "New Tirana" that had yet to come. It connected the city of everyday life to the ideal one. In terms of topography, the boulevard's alignment, almost parallel to the Dajti Mountain range, made visible the north–south extension of the Tirana valley, which is slightly inclined to the west and gradually descends through a sequence of river necklaces (e.g., the *Lana, Tirana,* and *Tërkuza* rivers). Visualizing these important geographic features at the city scale reinforced the identity and orientation of the city, mediating a dialog between the geographic and human scales.

Related to the axis as a "political diagram," as Kostov argues for other cases, Tirana's boulevard "celebrates monocentric domination" (2003, p. 174). In Tirana, this domination was guaranteed, not only because of the boulevard's clear contrast with the preexisting organic structure, but also because of the way it was inserted into the overall urban scheme of the future city, which emphasized its effects (Figs. 4.5 and 4.6). The boulevard became the generator axis for future developments based on a new and different logic. Interestingly, in the original design concept, King Zog's palace was to be located on the dominant hill at the south end of the axis (later substituted with the headquarters of the Fascist Party, or *La casa del Fascio*). Instead, the land along the axis, designed according to an orthogonal layout, was distributed to highly ranked clerks and ministers (Fig. 4.5b). It was a symmetric structure in which urban hierarchy was related to political status.

This new part of Tirana, rooted in the central axis, was a kind of "ideal city" next to the organic one. The ministry square, redesigned after Brasini by Florestano De Fausto in the neoclassic style, was the most imposing complex in the city—a new urban scale and public dimension in city life. The designed city was already a concrete reality, and it brought with it a new age in Tirana. Sitting, alien, alongside the organic city, this architectonic representation fulfilled its political goal: to legitimate the newly created nation, no longer part of the Ottoman Empire.

The urban and architectonic interventions in Tirana significantly intensified during the Italian Fascist occupation of 1939–1943. The main public buildings designed and built in Tirana during this period were strongly influenced by *Stile Littorio,* the Italian current of what Frampton (2007, pp. 204, 214–215) defined as the "New Tradition": a style of architecture developed in close relationship with political power. The most typical expression of the "New Tradition" in Tirana was the complex of *Piazza Littorio* (Fig. 4.7), which formed a quadratic square at the southern extreme of the main boulevard. This complex—which included the *La casa del Fascio* at the center of the axial perspective, the *Dopolavoro* (after-work activity complex), the stadium, and the gymnasium—was representative of the architectonic lexicon of *Stile Littorio,* from its stripped and purified classicism to its lithic monumentality and solidity. Covered in marble and travertine, it was originally supposed to be adorned with large statues, but these were never completed.

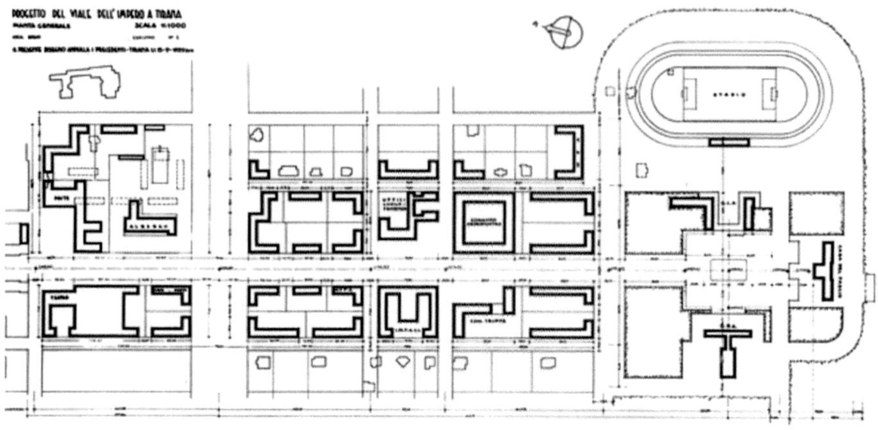

Fig. 4.7 *Viale dell' Impero* and *Piazza Littorio*, designed by Bossio and finalized by Lambertini and Poggi (NTA). Only the buildings of *Piazza Littorio* were realized

From an urban perspective, this new complex outside the existing city paralleled EUR in Rome (Universal Exhibition of Rome), but on a much smaller scale. Frampton's (2007, p. 215) description of EUR as a monumental utopia separated from reality is also true for Tirana. This new square, together with the boulevard (named by the Fascist administration *Viale dell' Impero*) and the ministry square, introduced a new dialectic and reinforced the tension between the organic and designed parts of Tirana (Figs. 4.8 and 4.9).

During the time of the monarchy up to the end of the Second World War, the southern part of Tirana on both sides of the new boulevard, designated by King Zog for the government clerks, became the ground for modern expressions in architecture, especially for housing and urban villas. While some urban villas were successfully adapted to the preexisting organic structure, avoiding the need for transformation within historical areas, in "New Tirana" they were used as foundation devices, shaping that part of the city. The orthogonal grid and prospected avenues, flanked by urban villas overlooking tree-lined open courtyards inspired by an Occidental model,

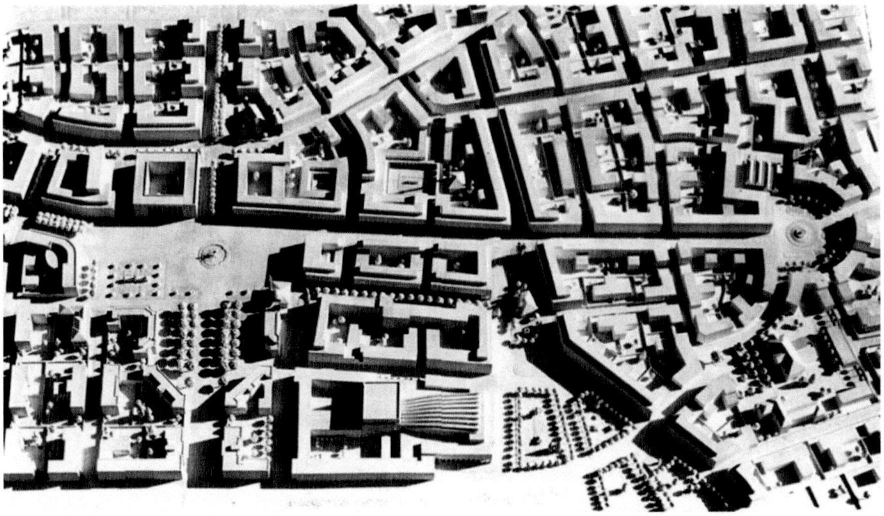

Fig. 4.8 Plan for the systematization of the ministry square and the Old Bazaar (NTA). This represents a new planning and design culture in opposition to the previous organic and Ottoman styles

contrasted radically with the organic fabric of the interlocked winding streets and closed perspectives rooted in Ottoman culture (Figs. 4.10 and 4.11). The owners of these villas held a specific social status in Tirana: the middle/small bourgeoisie. Most had been educated in Western countries and sought to change the feudal, oriental, and "backward" image of the country. If we refer to Rossi's (1984, pp. 77–82) statement that "the house represents the manifestation of a culture," it is clear that the urban villas built during this period contributed to the expression of a new (European) culture in Tirana, distinct from the organic (Ottoman) one. This typological difference is still visible in the present.

As a result of this progressive Europeanization over the 15–20 years to the end of the Second World War, two Tiranas (Fig. 4.9) with distinct urban and architectonic characteristics were created. The first continued to grow from the organic radials, while the second emerged from the re-foundational axis of the boulevard. However, as the efforts of these two decades focused primarily on physical structure and image rather than on education or other aspects of civil life, they could not fully give to Tirana the aspect of a European city. Despite these planned interventions and the appearance of the new public buildings, the organic fabric remained a refuge for the historical enzymes of the city. Nevertheless, parts of this fabric and some important monuments, such as the Old Bazaar, were damaged or demolished after the end of the Second World War. Thus, Tirana underwent the simultaneous clash and intertwining of drastically different cultures, as reflected in many aspects of daily life (e.g., the anthropology of space represented at once or separately the Ottoman and Western civilizations). After this period, Tirana was no longer the same. The architecture and

Fig. 4.9 Two Tiranas: the organic city and the axial–orthogonal designed city. From the closed perspectives of labyrinthine streets and dead ends to open and continuous perspectives; The boulevard as a generator for the restart of Tirana

urban traces from this period constitute the first base for Pattern 2: Recording Over analyzed in Chaps. 5 and 6.

4.1.3 Framing the City (Fig. 4.12)

Between 1944, when the Communist dictatorship was installed in Albania, and its fall in 1991, other important changes occurred in the structure of the city. In Albania, as in many other totalitarian regimes, architecture and urbanism were coercively allied with power. They became ideological instruments for the so-called progressive transformation of the country and the creation of a new physical and social reality. This reality was supposed to be nothing less than the ideal habitat for a new humankind, purified from luxurious claims and the vices of capitalist society. The real guiding principle for the reconstruction during this period can be seen in Eugène Pottier's (1871) Communist international text: "'Tis the eruption of the end. / Let's make a clean slate of the past, / Enslaved mass, arise, arise! / The world's foundation will

Fig. 4.10 Organic Tirana, rooted in Ottoman culture (NTA)

change, / We are nothing, now let's be all!" This propaganda was aimed at motivating the despotic imposition of the new city over the old one. This also signifies a clash between the embedded Ottoman and bourgeois mentality and the new Stalinist culture penetrating from the Soviet Union and other Eastern Bloc countries.

Fig. 4.11 **a** The project along the main axis of the boulevard according to the regulatory plan of 1939, redesigned based on the ideas of Kohler (NTA). **b–d** Urban villas built in Tirana during the same period

Despite the isolation that dominated this period, Albania followed a similar track to other Eastern Bloc countries in terms of architecture and urbanism, albeit within the framework of a struggling economy, a lack of technology and know-how, and aggressive ideological pressure and dogmatic irrationalities. This was generally reflected in the poor adaptation of the models of reference, often taken from other Eastern Bloc countries, especially the Soviet Union, Bulgaria, or Romania. The inertia of this influence lasted even after the breakdown of relations with the Eastern Bloc countries (1960) because Albania's administration and education systems were full of people educated in those countries.

Fig. 4.12 Collective residential blocks late 1960s/early 1970s; (NTA)

Regulatory urban plans based on zoning as the main regulation criteria were the most important instruments for large-scale urban transformations between 1960 and 1990. They were also political instruments aimed at linking territorial transformations with the ideological and political goals issued in the 5-year national development programs. In the regulatory plans, Tirana was considered a mechanical collection of pieces that could be centrally removed or dismantled. The idea was to dilute the historical city, to gradually sever all ties with the compromising and regressive capitalist past. In the "new society," bazaars, religious centers, urban villas, and many other artifacts were seen as evil (Fig. 4.13).

Clearly, the preference was for erasing the old neighborhoods rather than enhancing them. This erasure began with the most symbolically significant historical places and objects associated with the emergence and foundation of Tirana (Fig. 4.13): the Old Bazaar, Old Mosque, and *hamam*. It then focused on important commercial streets such as *Rruga e Barrikadave* and many residential areas of the organic period. Taking the place of these areas were modest housing blocks and ceremonial voids, celebrated by the construction of buildings to act as centers of the new political and cultural life.

Fig. 4.13 The historical center of Tirana was almost entirely erased. The Old Bazaar and the intense matrix were substituted with emptiness—the space of oppression (NTA)

The greatest transformation during this period came from the standardized collective 4–5-story apartment blocks that sprung up around the city. Designed according to a strict ascetic philosophy, these apartments were violently implanted within existing neighborhoods after cleaning up the footprint needed for the buildings and their associated access roads. In other cases, the main streets were "curtained," through the insertion of new apartment blocks creating deep wounds and morphological tensions within the organic urban fabric (Fig. 4.14). The infiltration of these alienating frames heavily damaged many parts of the city. No attempts were made to explore the binary relationship between the proposed housing typologies and historical urban morphology. Further, in most cases, the new blocks were never completed with the social services, playground equipment, and sports facilities proposed as the social catalyzers for the new collective life. Bulldozing parts of historical neighborhoods and sometimes single houses without consideration of the morphological organism created the conditions for what was to come in the post-dictatorship period.

The atmosphere that reigned within the faceless frames of these blocks was one of desolation and emptiness (Figs. 4.15, 4.16 and 4.17). This anonymous and faded city, deprived of human expression, lost any attractiveness that might come from the difference. The oppression of this environment had serious long-term consequences for Albanian society. The loss of identity and social depression that followed the collapse of the Communist regime can also be partly attributed to the alienating nature of the public spaces. Despite the claims of an equal society during the communist era, discrimination occurred from center to periphery, between cities, and within the Communist Party based on rank. The monocentric domination of the main axis of the boulevard and its absolute symmetric structure perfectly mirrored the hierarchical society of the Communist period. Emptied of the energy of the people, the city was becoming a metaphysical structure and scenography of power alongside the organic neighborhoods and anemic apartment blocks erected during this period.

The industrialization programs between 1960 and 1990 drastically changed the shape of Tirana. While the historical organic growth of Tirana during the nineteenth and twentieth centuries was generated by the development of the city as a market hub, growth after the Second World War stemmed from the development of the city as an industrial pole. During this period, the city stretched to the east, west, and north, in the directions of the newly built train station and expanded industrial zones. This expansion was accompanied by the construction of adjacent low-cost residential areas. However, with the exception of the textile factory complex "Stalin," which was built as a satellite outside the existing city (Fig. 4.18), the new developments did not extend more than 2–3 km from the historical center. The expanding structure of the city was also reconnected through a ring-radial road scheme, outlined in the regulatory plan of 1958, the first of the post–Second World War period. This plan, and that of 1989 (Fig. 4.19), further reinforced the monocentric features of Tirana (Fig. 4.20), based on the combination of radials and concentric rings. However, most of the operations proposed by these plans were never completed. In reality, Tirana was transformed internally by the insertion of distinct urban and architectonic typologies, and externally through new industrial extensions.

With the collapse of the Communist regime, it became clear that the dream of a "new city" in which everybody would be equal was no more. Instead, it was replaced by the desolation emanating from the poor housing blocks, the empty frames of which (Fig. 4.15) failed to provide the habitat for the proposed new model of life. However, they did create the conditions for their own later drastic transformation. These frames are the main base for Pattern 2: Recording Over analyzed in Chaps. 5 and 6.

Fig. 4.14 Cutting and framing the historical city (NTA)

Fig. 4.15 Hollowing the city: "social" spaces without social meaning. Instead of a collective life, they created the condition of emptiness within frames. (Top photo Driscoll; Bottommost photo from NTA)

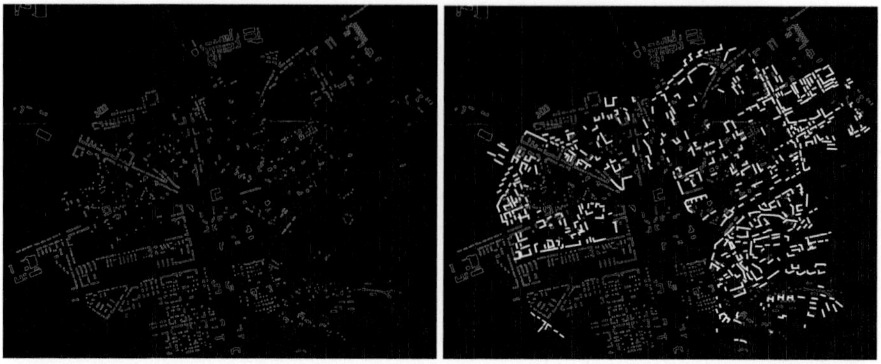

Fig. 4.16 New portrait of the hollowed Tirana: the central area of the city. On the left, 4–5-story apartment blocks, or frames of emptiness inserted up to the 1970s. On the right, interventions with the same typology up to the 1990s

The organic fabric, "pierced" by a monumental axis during the monarchy and Fascist periods, was now "hollowed" by the frames of emptiness. This was the third significant transformation in Tirana's urban morphology.

This plan reiterated the establishment of a third ring level, originally proposed in the plan of 1943, and the extension of the existing boulevard to include a new access road from the north. The periphery, mostly dedicated to economic activities, was confined by a green belt outside the third ring level as a limit to city growth. Conceived within a different sociopolitical system, most of the actions proposed by the plan were not realized.

4.1.4 Spontaneous Waves of Urbanization

The decay of the Communist utopia and the dramatic socioeconomic collapse that followed led to rapid and unprecedented urbanization in Tirana and Albania's other main urban areas. The new conditions drastically changed the role of the state, including with regard to construction activity and territorial control. The emergence of a shapeless peripheral city and the densification of the city center through massive infill developments and smaller parasitic structures, which gradually reduced public spaces, were completely novel phenomena in Tirana. The self-organized "informal city" gradually and almost imperceptibly became the reality (Fig. 4.21).

Efforts to improve living conditions in the central areas of the city, mainly through improving existing infrastructure and services, were followed by increased density along the main roads and infill developments, contributing to the loss of more public spaces. Moreover, freed from political oppression, claims related to former properties and pretentions about infringed rights, apparently lying dormant for decades, were resurrected. The only legal criteria restricting plot-based infill developments was

Fig. 4.17 Typical residential blocks in Tirana (early 1970s) soon after their completion (NTA)

Fig. 4.18 *Kombinat* city, a satellite neighborhood of Tirana, 6 km from the center of the city (established in the early 1950s; NTA). This center developed around the textile industrial complex "Stalin"

(in the best cases) the "respect" of a normative distance, measured geometrically and calculated arithmetically. The administrative illiteracy inherited from the period of a dictatorship made a holistic view of the ongoing phenomena impossible. The sheer size of the newly infiltrating developments, now 10–12 stories rather than 4–5 stories, made the situation even more dire. At the same time, people living in Communist-era collective apartment blocks began choosing to improve their housing

Fig. 4.19 The regulatory plan of 1989 (approved one year before the collapse of the dictatorship), authored by the National Planning Institute, Tirana; reproduced from Dhamo et al. (2016), based on NTA sources

conditions through extensions of small parasitic structures, further encroaching into public space. This phenomenon is further explored in Chaps. 5 and 6 under Pattern 2: Recording Over.

The second growth trend that appeared soon after the fall of the dictatorship was the self-development of the city outside the official boundary. The reform of land and housing privatization, starting in 1993,[2] was followed by considerable changes in the economic, social, and spatial structure of the city. As a result, self-organized developments emerged both peripheral to and distant from the original city limits in former agricultural fields distributed to farmers under law. This new part of the city extended in endless intricate tentacles, which wrapped around the preexisting rural settlements and former or abandoned state cooperatives and farms, resulting in an expanding and shapeless entity (Figs. 4.21 and 4.22). This phase of development was similar to the historical organic processes but of greater magnitude. Observations and analyses carried out in 2001 as part of the Strategic Plan for Greater Tirana found evidence for the massive expansion of the Greater Tirana area: from 12 km^2 before the collapse of the Communist dictatorship in 1991 to 32 km^2 in 1994 and 56 km^2 in 2001—an almost fivefold increase (PADCO et al. 2002, p. 1). This corresponded to an almost seven percent average annual population growth for the 1991–2001 period, Tirana's largest ever wave of urbanization. From a city of 250,000 inhabitants in 1991, Tirana reached 585,000 inhabitants in 2005 (based on municipality data).

[2] Two were the main laws guiding this reform: Law N. 7501 dated 19.07.1991 "On the land" privatization, and Law N. 7652 dated 23.12.1992 "On the privatization of state-owned housing".

Fig. 4.20 The space of oppression: images from the directions of the primary axes and radials entering the main ceremonial square (NTA)

Fig. 4.21 People did their best to survive—the "evaporation" process on the southeast periphery of Tirana

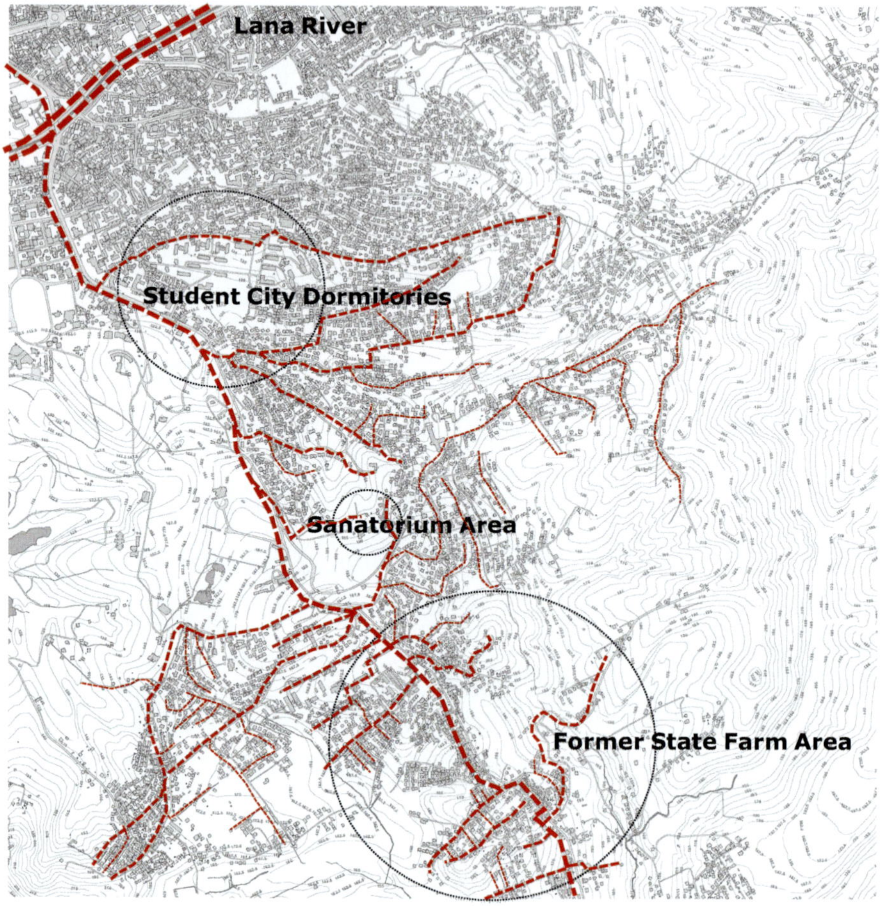

Fig. 4.22 The process of "evaporation" on the southeast periphery of Tirana. This process was driven by self-organizing informal and organic neighborhoods and resulted in an expanding and shapeless city

These newly self-organized residential areas were usually physically detached and morphologically distinct from the city center. Specifically, they were not compact, and there were often great voids between them. However, over the past decade, these areas have undergone organic densification and consolidation. This process goes through actions of micro-morphology transformation, adaptation, and incremental steps that as Jones says, occur from interactions within groups who manage to self-organize and co-evolve by reflecting changes in space, time and society (2020, p. 181). In the total lack of the top-down planning, the unique patterns of the resulting micro-morphology are related to the "process and type of micro-adaptation" of the housing, plots and blocks (2020, p. 182). This is mostly reflected in continuous redefinitions and modifications in the use of walls, windows, or encroachments generated by multiple iterations of the housing frontage setback, alignment, set forward, set above, or as this author calls it, "interface creep" (2020, p. 188, 189). As we will see in Chaps. 5 and 6, their morphological and other characteristics have been subject to their specific location and distance from the center.

One reason for the urban sprawl created by these self-organizing developments was the high demand for housing created by domestic rural migration, combined with the poor economic conditions and inability of the government to offer affordable housing alternatives. Other reasons included the lack of an operational and realistic planning system, such that the focus was on banning and prohibiting developments rather than enabling; a lack of clarity around land ownership; and a cultural attitude of disregard for the rule of law. Urban spread in Albania, especially in the first period after the collapse of the Communist dictatorship (1994–2000), started as a phenomenon of the poor. However, at a distance of almost three decades, the success in some areas of residents' strategy of incremental adaptation aimed at their economic and social recovery has been noticeable. During the past decade (2010–2020), sprawl has also emerged as a phenomenon of the rich, who have chosen to isolate themselves in gated communities on the periphery of the city.

Starting from the beginning of the new millennium, several attempts were made to create a vision at the metropolitan scale. The 2001 Strategic Plan for Greater Tirana, which was based on a classic and technical planning approach, proposed metropolitan governance for the Tirana region. The 2004 "Tirana Metropolis" was an inspiring strategic vision for the definition of a more "intelligible metropolitan geography" as an "archipelago" that was still under formation (Berlage Institute 2004, pp. 21, 22). Due to institutional disagreements and a lack of administrative capacity to absorb new ideas, no progress was made on implementing the proposals from these plans. Since 2010, the urbanization trend has calmed. A new plan for Tirana was approved in 2012; however, for political reasons, this plan was abolished and replaced with another plan in 2017. None of these plans have represented a paradigmatic change in the city's management. Rather, they have primarily focused on increasing the buildable ratios and intensities, including in the historical areas. Formulated under the "shining" propaganda of star architects, among others, these plans have served to legitimize hidden interests by giving them legal status and have increased social tensions related to urban issues. In addition, they opened the way for evictions from informal settlements, the residents of which left their homes only after protracted protests, fights, and, finally, police involvement. These outcomes, which stand in opposition to the reality of the city, are a clear sign that Tirana cannot be managed by traditional city planning instruments.

The most emblematic projects of this decade have included the extension of the 3 km northern Tirana boulevard and the organization of a 7 km riverside park, touted as giving the city a new dimension. The challenge for these projects has lain in how they approach the existing parts of the city. A considerable portion of the area they relate to contains informal settlements, which exist only in physical reality, not on the map. Unfortunately, the first signs (i.e., these projects' proposals) are not promising in this respect.

While much effort has been spent on the redesign of the central areas, especially those located along the main axis of the boulevard, the periphery and newly created suburbs have been neglected. The unwillingness to treat urban problems holistically has created a double standard: in the "city of appearance" legislation and other standards are supposed to apply, while in the "hidden city" these same standards are considered non-obligatory. This is because no further steps have been taken to create the legal conditions that would support affordable housing.

In the three decades that followed the collapse of the dictatorship, Tirana drastically changed from both within and without. These changes have affected its internal typological and social structure, its size, and its relationship with the suburbs and agricultural or natural surroundings. In the wake of the unsuccessful sterilization strategy of the proletarian city, naturalness once again invaded the city with full force.

As explained in Chap. 2, the different labels that came to define and characterize this new phenomenon of emerging and spreading informal neighborhoods (e.g., informal and illegal) are simplistic and hinder the genuine understanding of this phenomenon and its related urban effects. Generations and emergences formed under this condition are analyzed further under Pattern 2: Recording Over and Pattern 3: New Organic in Chaps. 5 and 6, as part of the natural city and with direct reference to the Historical Organic city. These two chapters explain the main processes and characteristics of these patterns, and how they can be understood from a holistic approach.

In conclusion, after having its organic fabric pierced and hollowed during the periods of the monarchy and dictatorships, since 1993 the city has been blurred by spontaneous waves of urbanization. This represents the fourth major transformation of Tirana's urban morphology.

4.2 Conclusions

This chapter highlights the role of the main sociocultural, political, and governance drivers in shaping public and private spaces in Tirana in relation to the main historical periods. Understanding these specifics in the generation, evolution, and transformation of the city, is the starting point to define the originality of the main urban patterns, and to analyze them as emerging from interrelationships within a larger spatial–temporal continuum, as seen theoretically in the previous chapter and will be seen more specifically in Chaps. 5 and 6.

Under the strong influence of the Ottoman occupation (from 1385 until 1912) initial nuclei and the first layer of the organic city (1614) emerged. The newly emerged *mëhalla* and the generation of many parts of the city were traced by an indelible imprint of the *imaret* system, which among others included religious complexes, schools, and baths, serving as pre-urban nuclei clustered around the mosque. The urban structure underpinned by this way of life created deep-rooted imprints, which, as we will see, will reemerge in future developments in Tirana.

King Zog who ruled Albania from 1922 to 1939 and his attempt to Europeanize Tirana trough some great works, as well as the very short period of Italian fascist occupation, from 1939 to 1943, by importing *Stile Littorio*, laid the first "artificial" layer in Tirana. This "second" Tirana represented in Fig. 4.9 can be interpreted as a period of absolute and authoritarian geometric planning and design and the beginning of a dichotomy that would be reinforced in the next period of dictatorship.

The 47-year period of dictatorship (from 1944 to 1991) and absolute self-isolation that followed the end of the Second World War was characterized by the preference for replacing the old neighborhoods with collective 4–5-story apartment blocks. This third significant transformation in Tirana's urban morphology resulted in the creation of the anonymous and alienated city ruled by desolation. This was the most significant imprint that inevitably drove inhabitants in acts of re-appropriation of public space. This phase was managed by rigid planning instruments considering the city as a mechanical collection of pieces.

The subsequent spontaneous tides that exploded soon after the collapse of the dictatorship and that have lasted for almost three decades have challenged the strict rules of the period of dictatorship. The high demand for housing was nourished by domestic rural migration, the poor economic conditions, and the inability of the government to offer affordable housing. The densification of the city center, and extensions of small parasitic structures encroaching into public space, as well as the emergence of a self-organized peripheral city, registered Tirana's largest ever wave of urbanization. This phase represents the most drastic transformation of Tirana's urban morphology. In this sense, this chapter is important to start understanding the impact of self-organized adaptive mechanisms occurring in the small-scale interactions and reflected in the micro-morphology (Jones 2020, p. 192). Recognizing this impact is crucial for complexity-based approaches in managing the evolution of cities (2020, p. 182).

As noted in this chapter, Tirana has been "processed" alternatingly by spontaneous waves of urbanization and authoritarian decisions. Despite these efforts throughout Tirana's history, it remains a natural city, where the evidence of authoritarian design and the attempted recording over of the people are clearly visible in many parts of the city and at different scales. The emergence of spontaneity, in a way, repaired the dichotomy inflicted by rigid planning and authoritarian design. As a result, Tirana exists in a context in which there are no clear limits or differences between fundamental categories (e.g., planned/unplanned, designed/un-designed), such that the geometrically designed city is organically deregulated (Figs. 4.23 and 4.24). These seemingly contradicting realities in fact constitute the specifics of Tirana; they complement each other and are part of the same whole.

Fig. 4.23 **a** The erosion of a Historical Organic area through infill densification with multistory collective apartment blocks. Gray buildings were added between 1945 and 1990 while black buildings were added between 1990 and date. **b** The process of Recording Over, mostly through infill low-rise developments in the free area framed by housing blocks built before the 1990s

The mental imprints forged during the historic periods presented in this chapter can be considered as wave factors transferable in spatial–temporal continuum, as explained in the previous chapter. They constitute the base from where the main motive underlying the pattern generation processes, the network of relationships and the respective qualities emerge. In this sense, the urban patterns are not physically limited entities. Generations and emergences formed under specific historic and

Fig. 4.24 No clear limits or differences between planned and unplanned. **a** an area considered informal, interspersed with some formal buildings. **b** The bottom image shows an area considered planned and designed, but still undergoing the process of densification

social conditions, as shown in this chapter, will be analyzed in the next chapter under Historical Organic, Recording Over, and New Organic patterns.

References

Aliaj B, Lulo K, Myftiu G (2003) Tirana the challenge of urban development. Co-Plan, Seda, Tiranë

Dhamo S, Thomai Gj, Aliaj B (2016) Tirana qyteti i munguar. Polis_Press the series of scientific publications of POLIS University, Tiranë

Frampton K (2007) Modern architecture—A critical history, 4th edn. Thames & Hudson world of art

Frashëri K (2004) Historia e Tiranës—Historia e Tiranës si qytet deri më 1920, vol 1. Botimet Toena

Hahn Von JG (2007) Studime Shqiptare. (The original title: Albanesische Studien. Translated by Dashi V, Koçi A); Instituti i Dialogut & Komunikimit, Tiranë

Ingersoll R, Kostov S (2013) World architecture—A cross-cultural history. Oxford University Press, New York, Oxford

Jones P (2020) The role of adaption in changing the micro-morphology of informal settlements. In: Di Raimo A, Lehmann S, Melis A (eds) Chapter 12 in the book Informality through Sustainability—Urban Informality Now. Routledge, London and New York

Kostov S (1995) A History of architecture—Settings and rituals. Oxford, New York
Kostov S (2003) The city shaped—Urban patterns and meanings through history. Fourth printing, Bulfinch Press AOL Time Warner Book Group, Boston, New York, London
Merriam Webster online vocabulary. https://www.merriam-webster.com/dictionary/pattern
PADCO, Value Add Management Services, Mix Technic (2002) Strategic Plan for Greater Tirana, vol 1. Working report under the Urban Land Management Program, financed by the World Bank and the Albanian Government
Pottier E (1871) Chants Révolutionnaires. Paris, Comité Pottier, [n.d. 1890–1900]. https://www.marxists.org/history/ussr/sounds/lyrics/international.htm
Rossi A (1984) The Architecture of the City. In: Rossi A, Eisenman P (eds) Revised for American under the series Oppositions Books. The MIT Press Cambridge, Massachusetts and London, England
The Berlage Institute (2004) Tirana Metropolis. Research report under the guidance of Zenghelis E, Aureli PV, and Tirana Summer Academy park projects

Legislation

Law N. 7501 dated 19.07.1991 On the land
Law N. 7652 dated 23.12.1992 On the privatization of state-owned housing

Chapter 5
Tirana Patterns at a Glance

Abstract In this chapter, I present an overview of Tirana's urban patterns. For each of the three patterns, Historical Organic, Recording Over, and New Organic, I present the main interacting forces/actors in their historical social context, and highlighted the correspondence between the essential quality of space and the social substance. For this, representative areas are selected, observed, and analyzed in relation to their site-specific emergent quality and formal/informal governance relationships underpinning the self-organized process based on a set of principles and arrangements. Illustrations include maps, synthetic diagrams, and pictures reflecting on the historical or current conditions for each of the patterns, starting from the city to the local scale. Patterns at a glance, is a preliminary step toward a more holistic understanding of urban patterns as emergent phenomena, a subject that will be elaborated into more detail in the next chapter.

Keywords Urban patterns · Historical organic · Recording over · New organic · Emergent · Generating principles

In the previous chapter, I outlined the main characteristics of Tirana in relation to the main transformative events. By investigating the city as a historical process, we can better understand whether specific urban conditions are the result of certain recognizable patterns that emerge from individual and societal behaviors or interactions between them. Thus, based on the specific conditions highlighted in the previous chapter, three main generative patterns were identified: Historical Organic, Recording Over, and New Organic.

In this chapter, I give a general overview of the urban patterns in their historical and social context, as well as of the main forces that motivate and set in motion the social structure that generates these patterns. In Chap. 3, I gave some general concepts about urban patterns, among others, as being the capacity to self-organize in complex behavior, or as owning site-specific emergent qualities resulting from underlying structures of interrelationships; and as containing a structure of formal/informal governance relationships based on group affinity. Based on these concepts and the conclusions of the previous chapter, here I refocus on the specific urban patterns of

S. Dhamo, *Understanding Emergent Urbanism*, The Urban Book Series,
https://doi.org/10.1007/978-3-030-82731-1_5

Tirana analyzing each of them through representative areas and illustrating through maps, figures, and synthetic diagrams. For this, in the first section, I give some criteria for the selection of the representative areas. Then I argue how the qualities of space and the generating principles characterizing the analyzed areas are intimately related to the social substance and to the relationships based on group affinity. I provide indications of the dynamics and reasons behind these processes. Patterns at a glance is a preliminary step toward a more holistic understanding of urban patterns as emergent phenomena, a subject that will be elaborated into more detail in the next chapter.

5.1 Patterns Representative Areas in Tirana (Fig. 5.1)

The previous chapter helped to identify three main pattern generation processes: Historical Organic as generations that emerged with Tirana's foundation as an organic settlement at the beginning of the seventeenth century under the strong influence of the Ottoman Empire; Recording over as generations that emerged soon after the collapse of the regime in the 1990s, intimately related with the re-appropriation of the alienated/collectivized urban space in those parts of the city built during the dictatorship; New Organic as generations that emerged in periphery of the city or

Fig. 5.1 The selected areas showing the three main patterns, to be further analyzed: Historical Organic (1.1–1.3); Recording Over (2.1–2.3); New Organic (3.1, 3.2) (Google Earth)

on agricultural land, originating with the collapse of dictatorship at the beginning of 1990s, driven by the impoverished economy and rural migration.

These patterns are embedded in the specifics of the Tirana generation process as an organic city and its modality of growth in different moments of the history, and as we will see in more detail in this chapter, they reflect specifics in their self-organization processes and the corresponding morphological aspects. We already mentioned in Chap. 3 the crucial role of patterns in understanding the self-organized processes in urban analyses, and how these processes are materialized in morphological expressions composed of legible formal sequences that create systems of relationships. In this regard for each of the patterns, representative areas have been selected for further analysis and illustrative purposes. Below I give some criteria for the selection of these areas.

This selection was preceded by historical analysis and examination of specific morphological aspects of the whole city concerning historical and sociocultural shifts, which are reflected in the family structure, the way of living, the relation with society, interaction of different groups, and their relation with government structures. The consideration of these elements that drive to specific patterns is based on the knowledge of the specific reality of Tirana as drawn from the bibliography cited in the previous chapter, field observations over the last three decades by the author, discussions with other experts in the field, intuitions, subjectivity and the senses.

Further, the selection of specific areas was informed by the combination of the following criteria such as clarity in the legibility of the formal structure; commonalities or repetitiveness of the formal sequences; typology of the elements that stay at the base of the formal structure and their combination in the system of elements; hierarchical organization of these elements and the system of relationships between different levels; human motives and the underlying structures of interrelationships in the self-organized process; and actors involved under specific historical and social conditions that create the range of formal/informal arrangements at the individual, group or social level. In addition, priority has been given to those areas where the morphological aspects of the patterns still survive despite the damages of the urban fabric. Table 5.1 summarizes the selected representative areas, highlighting their main characteristics. As mentioned in Chap. 1, the investigation process uses a mixed qualitative and quantitative method (as shown in Appendix 1, the quantitative analyses are applied in one of the selected areas for illustrative purposes).

Thus, by analyzing patterns that emerged within a specific historical and social context, we can reveal the motives and ordering principles underlying the human relations that give meaning to specific places. In this sense, patterns are *par excellence* the entities imbued with the genes and meanings to be understood and potentially use in the planning and design process.

Table 5.1 Patterns and representative areas to be further analyzed

PATTERNS	Discernable characteristics and current status	Representative areas
PATTERN 1: **Historical Organic**		
Started from the beginning of the seventeenth century and continued in its original form up to the beginning of the third decade of the last century. Represents the initial nuclei and the first layer of the self-generated city The generation process combined spontaneity with the *imaret/kulliye* system (Ingersoll and Kostov 2013, p. 439), a foundational device common in the Ottoman Empire Development took advantage of the excellent qualities of the geographic container Social conditions were characterized by spontaneous improvisations	Intricate mesh of streets. An urban maze with hidden internal parts but with clear logic and structure Areas representing this pattern continue into the present. Despite systematic intrusions, these areas remain capable of influencing the course of "artificial" technocratic planning. In fact, the incongruous high-rise interventions in these areas appear trapped in an organic maze, at risk of being decomposed over time	The three selected areas below represent similarities and differences according to their locations in relation to the city center (Fig. 5.1): Area 1.1 *Mëhalla Dibrane* in Tirana North, central location (started from the beginning of the eighteenth century); Area 1.2 *Mëhalla* near Red Hill (started from the beginning of the eighteenth century); Area 1.3 *Mëhalla Dibrane* in Tirana East, peripheral location (started from the beginning of the twentieth century)
PATTERN 2: **Recording Over**		
Begins with the fall of the dictatorship in 1990 and continues nowadays at a slower pace. Contains the collage and sometimes osmosis between the geometric housing frames built during the periods of dictatorship and the layers spontaneously recorded over by people The generation process combined formal and informal spontaneous individual arrangements, organically parasitic or infilling the emptiness of the planned city Development took advantage of the emptiness and meaninglessness of the urban spaces between frames Social conditions were characterized by post-oppression trauma and a lack of public control	Geometrically ordered from the outside or along the main lines, but labyrinthine or blurred within In the most evolved cases, it closed the space from within and replaced the original geometric communication with a labyrinthic one The re-appropriative and humanizing process through this pattern continues to evolve. This process accelerates or slows depending on the level of control by authorities	The representative area is an urban transect (Fig. 5.1) with three sequences (starting from the center areas 2.1–2.3): *Shallvare* (built in the late 1950s), *Puna,* and *Gjykata* (built between the late 1960s and the early 1970s). It starts from the center and extends to what was considered peripheral before the 1990s The three housing complexes represent the full range of issues encompassed by this pattern

(continued)

Table 5.1 (continued)

PATTERNS	Discernable characteristics and current status	Representative areas
PATTERN 3: **New Organic**		
Begins with the fall of the dictatorship in 1990 and still continues nowadays at a slower pace. Represents the recent layers of the emergent city that dissolved the city form, and which manifest different characteristics depending on their location nearer the center or on the periphery The process combines individual arrangements in a spontaneous bottom-up self-organized settling process, which emerged over a relatively short period Development takes advantage of the preexisting poles of attraction (former villages, economic zones, and natural resources) and an unclear legislative framework Social conditions are characterized by post-oppression and economic shock and a lack of public control	Appears shapeless and unstructured but is based on self-organized emergent complex behavior, aimed at guaranteeing survival Currently, these areas are more consolidated (and are sometimes even of good quality) in the nearest periphery, and they are more scattered and closely aligned with agricultural activities or natural landscapes in the farthest periphery	The two selected areas represent the typical characteristics of this pattern based on distance from the center (Fig. 5.1): 3.1 *Pallati sportit*—Former aviation area (in the nearer periphery); first shelters emerged in 1993; 3.2 *Laknas*— Former agricultural area (in the farthest periphery); first shelters emerged in 1994

5.2 Pattern #1: Historical Organic

5.2.1 The Essence of the Pattern

This pattern emerged as the first layer of the organic Tirana. It is a direct descendent of the *imaret/kulliye* system that, combined with urban and social improvisations, constituted the main principle for the essence of the city-form. In fact, the evolution of the Historical Organic patterns in Tirana followed the same path up to the beginning of the third decade of the last century because it was well rooted in the specific model of living. As such, it played the most important role in the irregular city-shaping of Tirana.

The Historical Organic pattern is characterized by specific principles that aim to close and to predominantly experience the space internally, or from within. Based on these principles, the self-organizing process generated a visceral quality of space at different levels of the city hierarchy. Visceral, which is the keyword for the created space, represents the essential quality of the Historical Organic pattern. Given the context of the Ottoman occupation (already explained in Chap. 4), the push towards closing was due to the "Muslim's right to visual privacy" (Kostov 2003, p. 63) deriving from the principles of the "traditional Islamic city…" (pp. 62–63). Thus, *visceral* in this context means an "interiorized" neighborhood or cluster of houses aiming to close in or conceal itself. In this sense, it represents also a specific way of privatizing space and a passage from the urban or common to the individual. The visceral quality of space that was not a usual enclosure was an implicit requirement of the specific culture based on the patriarchal enlarged family model, where many generations of descendants were living together in a big house or in a nucleus created by several houses built within the same family plot. Some Albanian authors such as Muka (2001) and Shkodra (1984) described the nuclear-family model that arrived up to the beginning of the last century.

This conception of life, which was based on the right to visual privacy and varied through levels of permissibility, no permissibility, or even total prohibition of watching, served as an instrument for concealment that processed and shaped the entire livable space by refracting visual lines. This motive urged frequent and repetitive actions, such as deviating from the straight line in order to avoid urban vistas, isolating and wrapping inwardly and outwardly empty interstitials, closing rather than opening, or blinding unreachable areas through *cul-de-sac* solutions (Figs. 5.2 and 5.3). Interestingly, the literal Albanian translation of the term (*qorre*) for these types of streets is "blinded streets." This "hide and seek" play based on religious morality principles, which is at the base of these kinds of neighborhoods, dictated an introverted behavioral character, one directly reflected in the morphological expression of the neighborhood and the sequential closing of Eden(s) within Eden(s) (see also Fig. 6.8 showing *cul-de-sacs*).

In these kinds of neighborhoods, people act and do things based on perceived consent and without having prescriptive written regulations or a predesigned scheme to handle the settling process. The principle of perceived consent based on the

Fig. 5.2 Still existing *cul-de-sacs,* or dead-ends in the analyzed areas

"accepted social behavior" (Kostov 2003, p. 63) demonstrates the emergence of informal governance arrangements (Suhartini and Jones 2019, p. 23), applied in the space subdivision for public or private purposes at the neighborhood and lower levels. These arrangements are an informal interplay between different stakeholders, mainly residents, individually or in groups, which have their roots in "local norms, values and structures." often associated with migratory groups, using and adapting over

Fig. 5.3 Space configuration that regulates the behavior through neighbor's watching

time customary practices (Suhartini and Jones 2019, p. 29). These informal arrangements "gain their validity from socio-cultural orders" and connections to local tribes or kinships and their gradual adaption in the urban settings (Jones 2016 in Suhartini and Jones 2019, p. 29). In our case, local residents grouped on the patriarchal basis of extended family, tribe, or origin, and religious/missionary leaders who founded and disseminated the *imaret/kullie* system as a waqf institution, were the main actors in setting the principles and translating them into informal arrangements. As explained in Chap. 4, the public buildings, part of *imaret/kulliye* system, served as the center of well-defined *mëhalla*, which gradually evolved and transformed through continuous adaptations, while still preserving the visceral quality of space of the Historical Organic pattern, because of these formal/informal arrangements.

According to this logic and the range of arrangements, the space subdivision began by occupying the personal private space. In this sense, the aggregation starts inwardly. The public space was less important and was the result of what was left, mainly access roads. Thus, the public space was continuously redefined based on negotiations, and a *cul-de-sac* completing the full internal protection was the common solution, one where priority was given to more consolidated structures (Kostov 2003, p. 63). In this sense, Historical Organic is an interdependent, interlocked, and cohesive neighborhood where the comportment of individuals is based on negotiation and implicit permission (Fig. 5.4).

Within this system, the interdependence and coordination between adjacent structures were required for repairs and other issues, such as the construction of partition walls, or for the maintenance of the *cul-de-sac* that by tradition was considered to be the common property of the houses using it. Coordination was needed also during the plot densification phase, the passage from single-house plot to composed residential plot. Considering that each family consisted of more than one familiar nucleus, the distribution logic within the family plot itself was subject to the same principle—that is, respect for the privacy of each individual house within the family plot. Hence, lower-level paths could be drawn and redrawn continuously to provide access to the new houses and improve connectivity. In some cases, access was provided by opening doors in the confining partition walls in order to use the courtyards as private connection spaces and create easements. In most cases, the houses in this kind of neighborhood had more than one means of access, and this was used as an alternative circulation network only by the residents living there. In case of tension between two neighbors, the doors were closed.

This process required perpetual and endless modifications not only to the plot subdivisions but also to the degree to which internal and external spaces could be accessed. Even at this scale, the aggregation of plots and buildings aimed at the creation of the "private Eden," a physical and social equilibrium based on interactions, negotiations, and affinities as the base of an informally established social formula, or an "accepted social behavior" (Kostov 2003, p. 63). The capacity to adapt, typical for the natural city, is remarkable in this pattern. The concept of the visceral quality of space will be further explored and reformulated in the next chapter.

What I have argued so far is directly reflected in the generating principles of *mëhalla* that are intimately related to the social substance and to relationships based

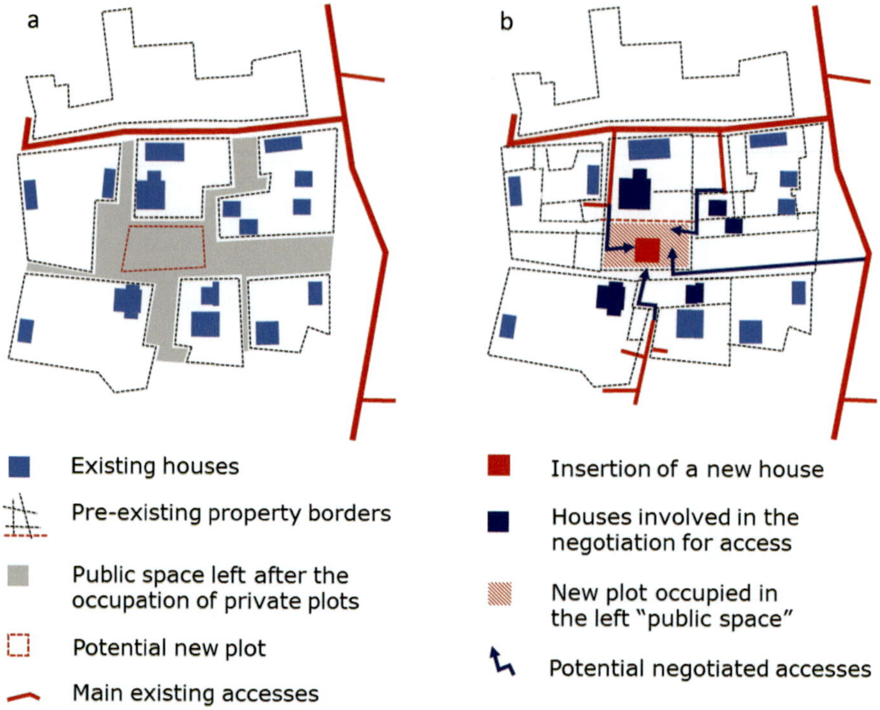

Fig. 5.4 **a** Inward aggregation of houses within private plots and redefinition of the left "public space" based on *cul-de-sac* solution. **b** Negotiations for wall partitions through micro adaptations. In each option, a portion of the garden is transformed into an easement (a semi-private passageway). In case of conflicts, access may be blocked or opened on one side or the other. This interlock could only be managed in the framework of an accepted social formula agreed between residents, individually and/or in groups

on group affinity. Their organization from the initial clusters to the entire neighborhoods appeared to be sporadic and irregular but in fact embodied a long-term and clear social meaning based on these principles.

5.2.2 *Brief Description of the Representative Areas*

Mëhalla Dibrane (area 1.1) in Tirana North was located in close proximity to the old bazaar (Figs. 5.5, 5.6, 5.7 and 5.8). The modality of growth was therefore very strongly influenced by the circular string lines generated by the bazaar, especially the 4th, 5th, 6th, 7th, and 8th (Figs. 5.6 and 5.7). They generated a perturbed and curved net where all the fabric of the neighborhood was woven. The lower level branched paths and the directions of refraction and bifurcation followed that net,

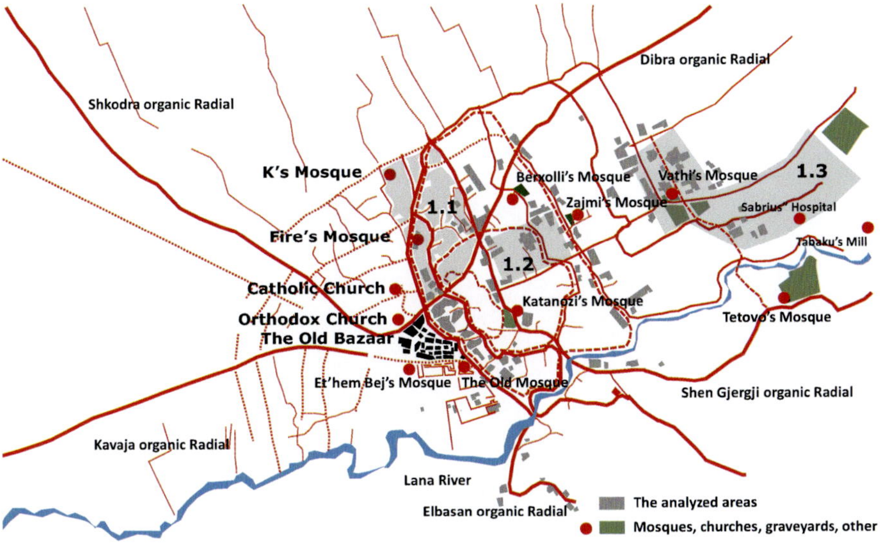

Fig. 5.5 Analyzed areas in relation to the main generative paths (Dibra, Elbasani, St. Gjergji, Shkodra, and Kavaja), the old bazaar, the mosques, graveyards, etc. as of the first decade of the last century

which, combined with other principles factors mentioned above, evolved and influenced the entire morphological character of the neighborhood. The isolation and the required privacy at the smaller scale were guaranteed by the meandering of paths or the aggregation of the houses following the deviating principle consisting in refractions and bifurcations or closing dead-ends at the level of groups based on affiliations (Fig. 5.9). In Chap. 7 and Appendix 1, this area is subject to an accurate empirical observation and measurement process followed by data registrations and other statistical calculations that will be explained more in detail in the respective chapter.

The "Read Hill" *Mëhalla* (area 1.2), being located near to the center, is mostly influenced by the old bazaar as a central element (Figs. 5.5, 5.6, 5.7 and 5.8). Four of the string lines, the 6th, 7th, 8th, and 9th (Figs. 5.6 and 5.7), rolling outwardly, influenced the modality of growth. While the 6th and 9th lines just confine this *mëhalla* and maintain a longer continuity throughout the city, the 7th and 8th lines, which are meandering and segmented, have a greater influence in defining the morphological character of the area. The logic that guarantees the required privacy, which is very similar to that of the previous case, is based upon aggregation and refraction at a smaller scale. For example, there are at least six bifurcation points from the 8th meandering line (Fig. 5.7). In addition, emanating from each bifurcation are at least two or three consecutive deviating levels that feed access to houses or to the cluster of houses (dead ends). Deviations follow previous land arrangements and ownership, but they are also influenced by the inward instinct to guarantee privacy

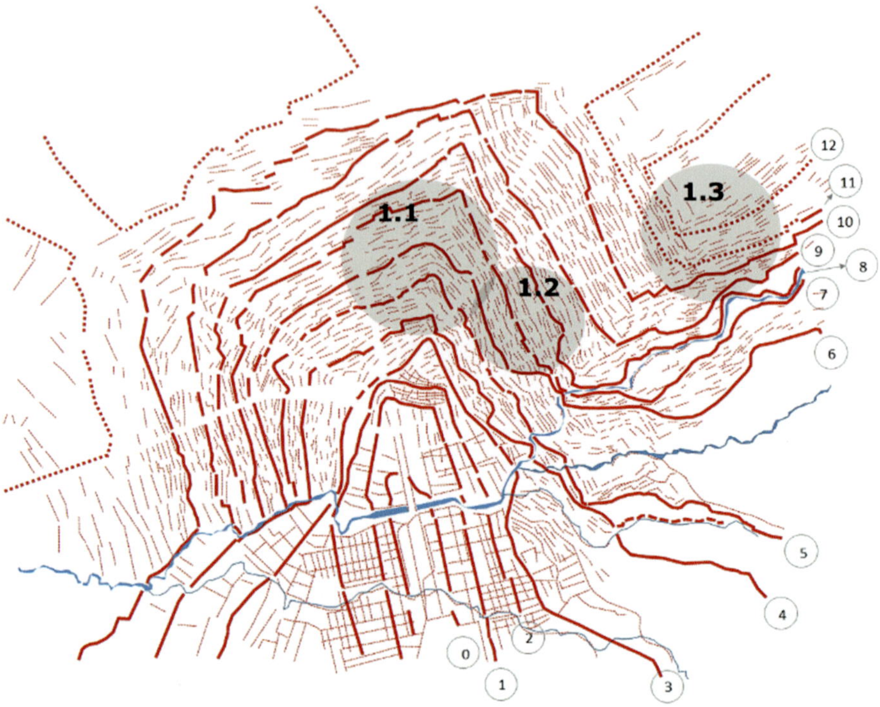

Fig. 5.6 Analyzed areas in relation to the modality of growth generated by the "gravity" of the old bazaar as a central element, expressed in the circular string lines rolling outwardly

at the personal, family, cluster, and neighborhood levels (Fig. 5.10). The dead-ends branch without reaching or intersecting any of the main paths bordering the *mëhalla.*

Mëhalla Dibrane (area 1.3) in Tirana East was located in the periphery (Figs. 5.5, 5.6, 5.7 and 5.8), between the 8th and the 12th string lines from the center (Figs. 5.6 and 5.7), and thus the modality of growth was marginally influenced by the old bazaar. At such a distance, the circular lines transformed into linear stripes due both to the lack of gravity from the center and the topography of the area. The neighborhood itself grew out of a path that deviates from the straight line of the main road, creating as such the premises for the privacy of the individual families and the entire group of families (Fig. 5.11). In addition to the principles already mentioned, also influencing this positioning and subsequent evolution process were the proximity to Lana stream, the moderate slope downhill to the river (appropriate for the drainage and wastewaters), the shallow area that connected the two slopes of the river, and the orientation of land subdivisions relative to the river's topography.

The series of the following figures give some additional information about the selected areas representing the Historical Organic pattern in relation to their transformation, insertion of different typologies, and the visceral enclosures (Figs. 5.12, 5.13, 5.14, 5.15, 5.16, 5.17, 5.18, 5.19, 5.20 and 5.21).

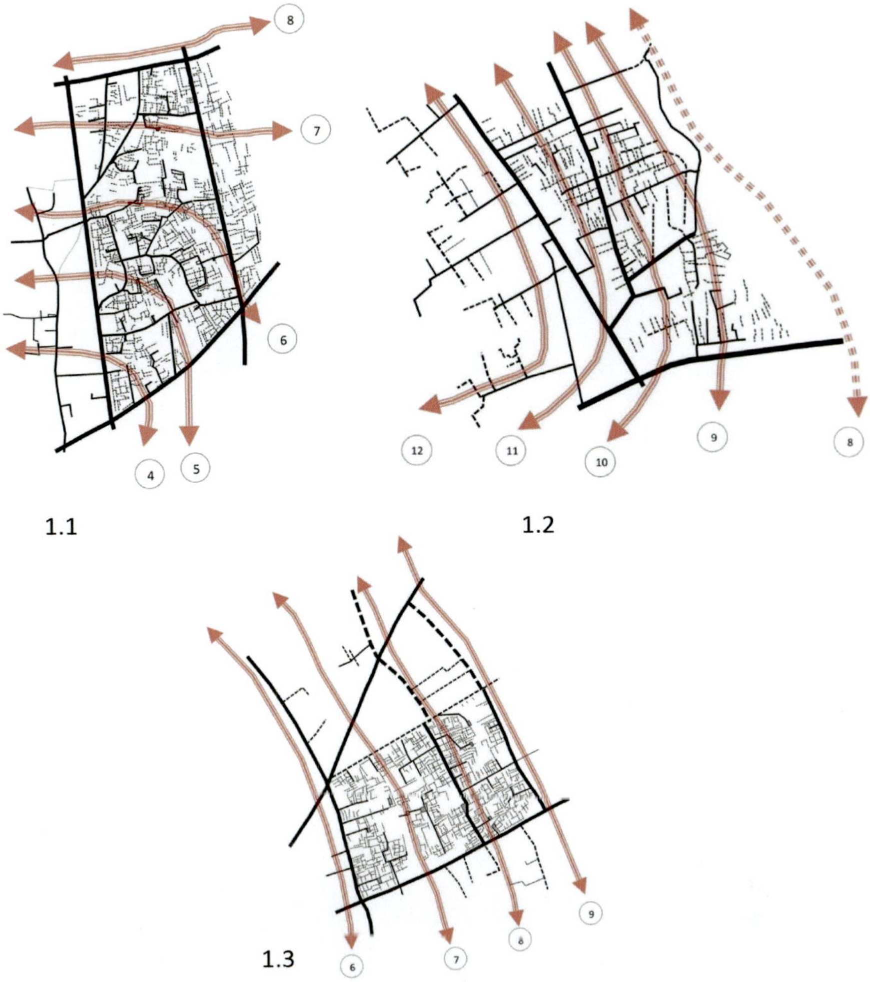

Fig. 5.7 Influence of the circular and linear string lines in each of the analyzed areas

5.2.3 The Simulacrum of the Historical Organic

Following changes in the traditional social structure after the Second World War, and especially after the collapse of the communist regime, the original principles which gave meaning to the organic patterns were drastically transformed. However, from the formal point of view, the new developments followed a similar logic as the historical developments. This was due to the physical and psychological effect of the permanencies from the previous frame that guided those developments. As mentioned in Chap. 3, we considered these visible–invisible effects under the notion of event and event horizons.

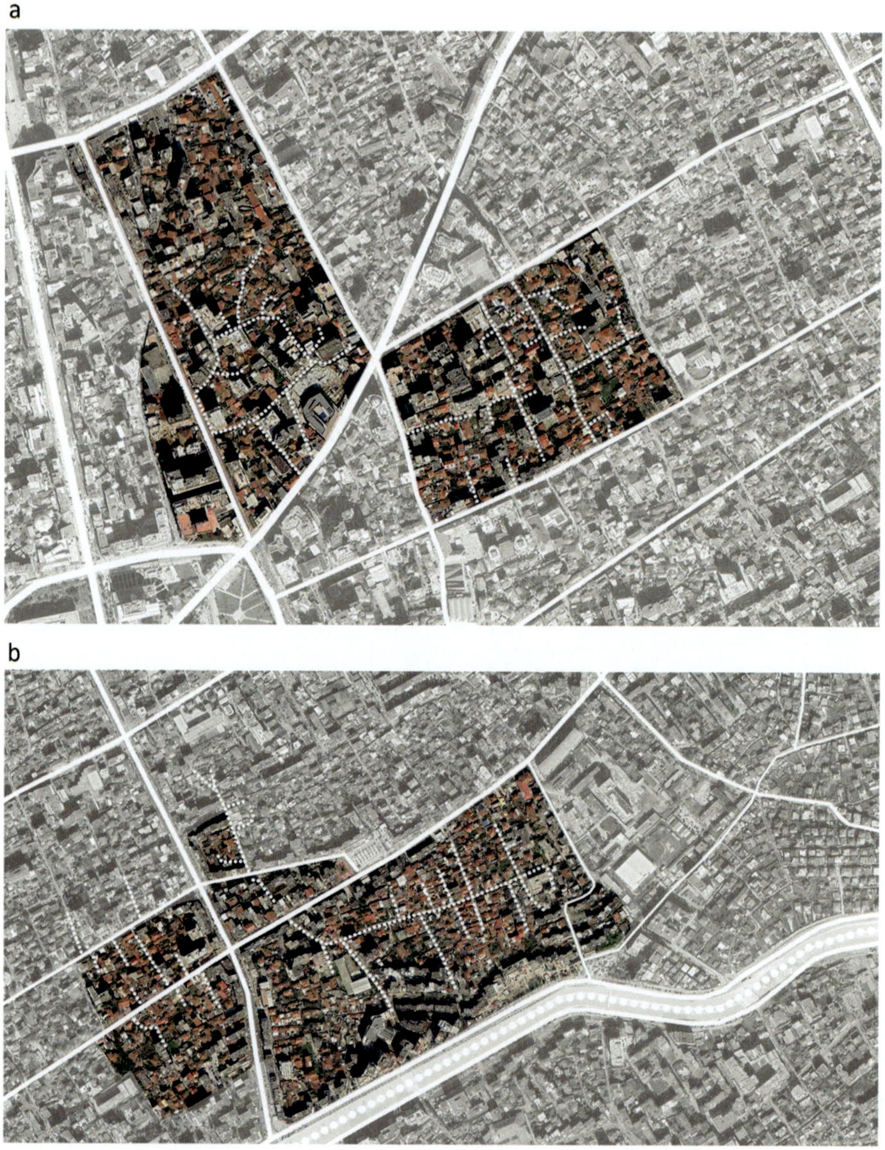

Fig. 5.8 **a** From the left to the right, *Mëhalla Dibrane* North (area 1.1) and "Read Hill" *Mëhalla* (area 1.2); **b** *Mëhalla Dibrane East* (area 1.3); See also Fig. 5.1 showing these areas on the map of Tirana

Fig. 5.9 **a** Social structure: groups of houses aggregated based on affiliation and relationship (clans, tribes, or large families); The base for traditional governance, local norms, and values. **b** The visceral quality of space. The wrapping process in relation to two "secret" visceral cores as the most hidden parts of the neighborhood. The process around the upper core is still going on (designed on the map of 1937 as the last moment the city arrived in its organic situation)

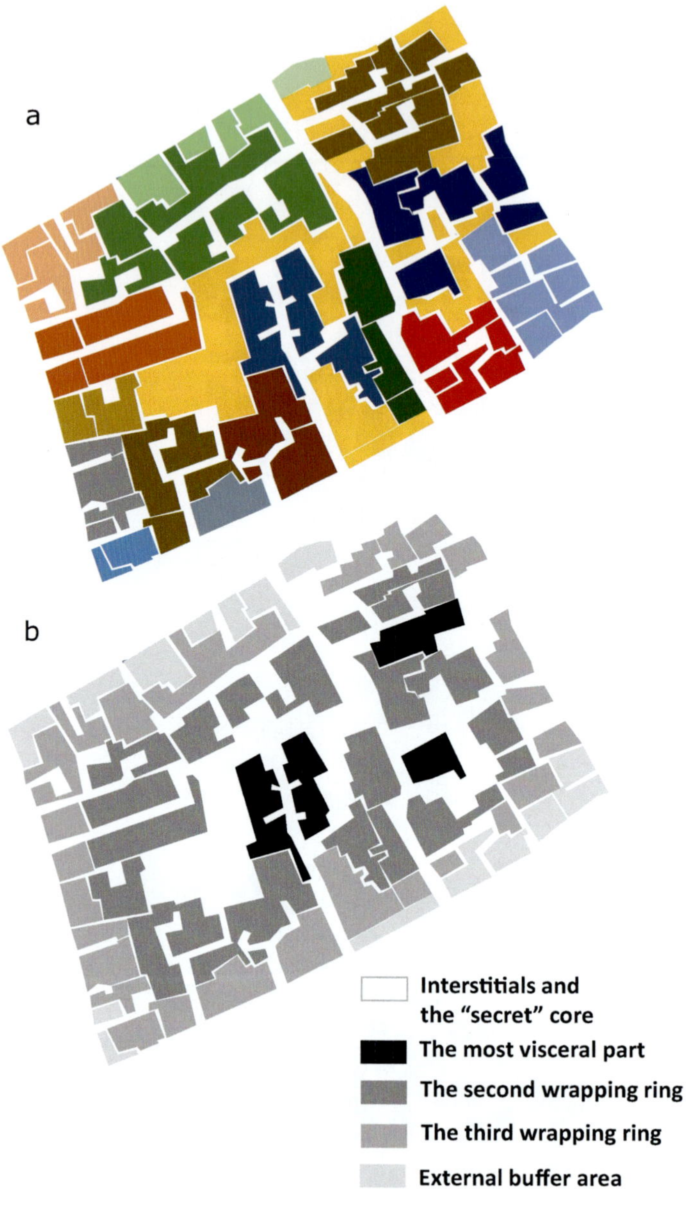

Fig. 5.10 **a** Social structure: groups of houses aggregated based on affiliation and relationship as the base for traditional governance. **b** Visceral quality of space. The wrapping process in relation to three visceral cores. The process is more consolidated around the central core and is still in progress around the two other cores (designed on the map of 1937)

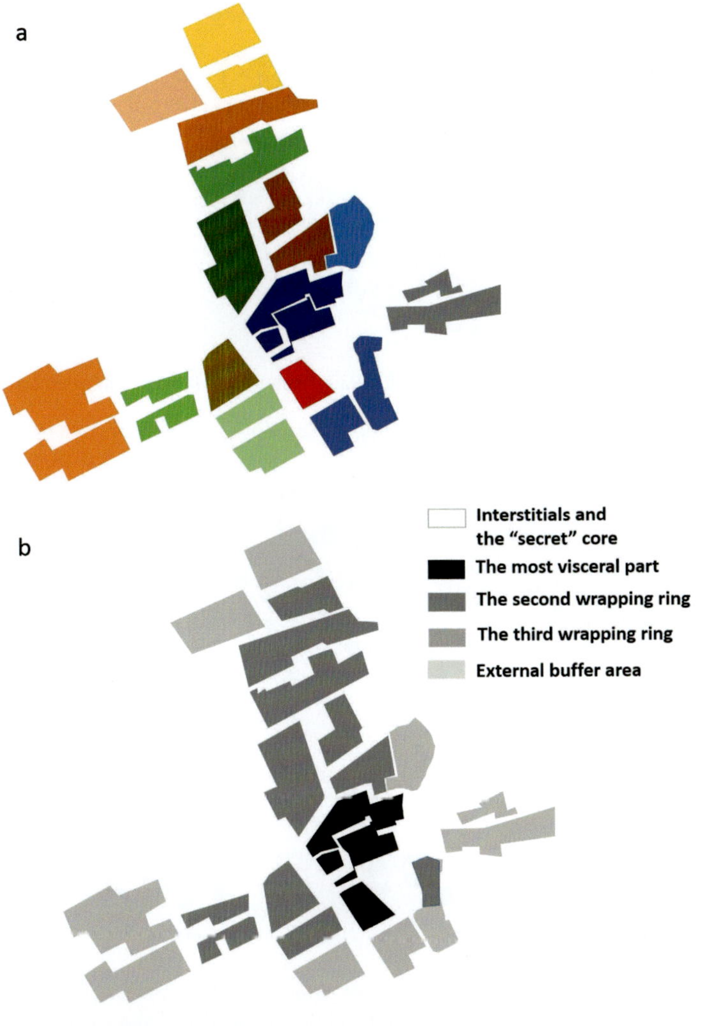

Fig. 5.11 **a** Social structure: groups of houses aggregated based on affiliation and relationship as the base for traditional governance. **b** The wrapping process in relation to the core in an almost radial modality from the center of the neighborhood. Together with the external buffer area, the housing groups located in combination with parallel interstitials along the radial axis reinforce the visceral quality of the central core

The interstitial free spaces characterizing these areas became increasingly densified in the subsequent periods. The real densification process started after the Second World War, when four- to five-story collective apartment blocks were introduced in the area, and it intensified after the collapse of the communist regime, especially in the central areas. In fact, the increase in land values put them under pressure and questioned their existence. Interestingly, as shown in Figs. 5.14, 5.17, and 5.20, the

Fig. 5.12 The analyzed area in 1921 (left), in 1937 (center), and the remaining urban maze to the present day (right)

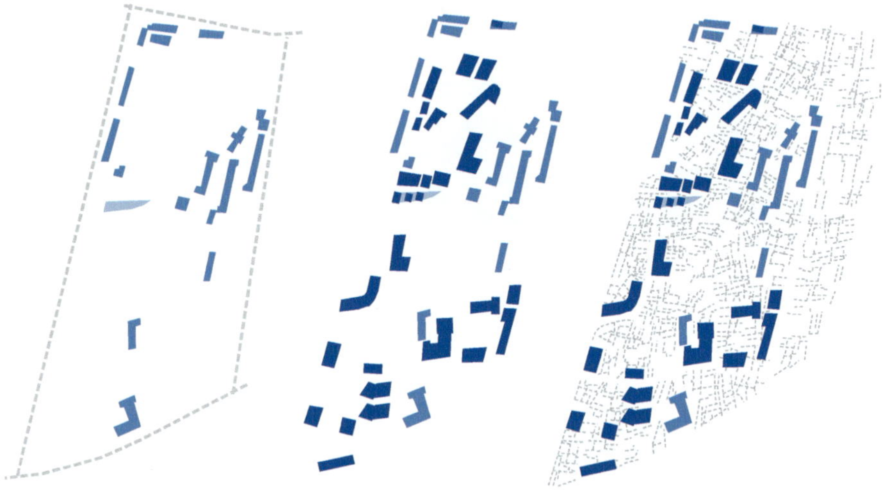

Fig. 5.13 The collective apartment blocks inserted in the area until the 1990s (left), the interventions added from the beginning of the 1990s until the present (demarcated at the center with a darker color), and the persistence of the hidden frame influencing developments (right)

interstitials between the historic developments and the external buffer area led to the ultimate enclosure of the area from outside through new developments, thus reinforcing the visceral quality through further wrapping. This happened even along the lower-level paths.

In some cases, parts of the historic structure were substituted with new high-rise typologies that drastically differ from the historic ones. However, their layout and the servicing systems were conditioned by the existing morphological system. They seem to be restrained within prior urban arrangements and property patterns, but on a different scale and a different social context. There are also some other

Fig. 5.14 The internal "secret" core (demarcated at the left with a darker color) and the external buffer (situation in 1937), the relationship of the empty areas with the houses (center), and the relationship with the collective apartment blocks (right)

Fig. 5.15 The analyzed area in 1921 (left), in 1937 (center), and the remaining urban maze to the present day (right)

Fig. 5.16 The collective apartment blocks inserted in the area until the 1990s (left), the interventions added from the beginning of the 1990s until the present (demarcated at the center with a darker color), and the persistence of the hidden frame influencing developments (right)

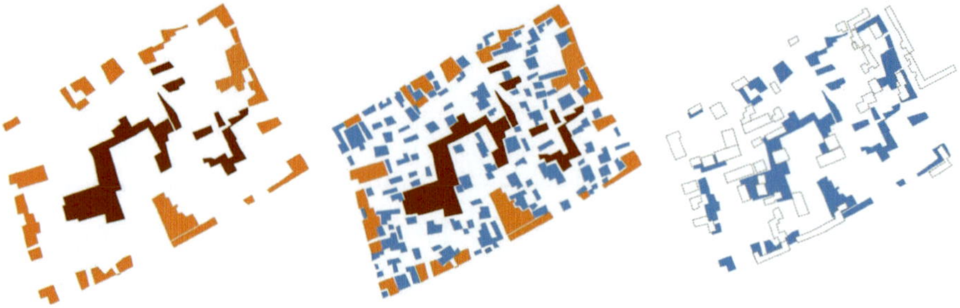

Fig. 5.17 The internal "secret" core (demarcated at the left with a darker color) and the external buffer (situation in 1937), the relationship of the empty areas with the houses (center), and the relationship with the collective apartment blocks (right)

Fig. 5.18 The analyzed area in 1921 (left), in 1937 (center), and the remaining urban maze to the present day (right)

Fig. 5.19 The collective apartment blocks inserted in the area until the 1990s (left), the interventions added from the beginning of the 1990s until the present (demarcated at the center with a darker color), and the persistence of the hidden frame influencing developments (right)

Fig. 5.20 The internal "secret" core (demarcated at the left with a darker color) and the external buffer (situation in 1937), the relationship of the empty areas with the houses (center), and the relationship with the collective apartment blocks (right)

correspondences with the preexisting layers, or latencies, including the behavioral pattern of the residents. Strangely enough, the principles of closing and deviating the space remained there even after the densification and intensification process. They were manifested in the highly complicated interlocked structures and the dysfunctional intricate access to the new buildings. In addition, the parterre "Private Edens" transferred to the top of the buildings, taking the form of "flying villas," which are normally occupied by the former landowners. The logic of the Private Eden sometimes is manifested in the high walls that enclose protected oases for children or for other semi-public or private activities that are formally or informally secluded. These principles, embedded in the hidden frames of prior arrangements, and in the cognitive spaces, co-evolved in the framework of a wider range of formal/informal actors and managed to survive.

Therefore, from a morphological point of view, the analyzed areas currently represent a collage that contains the co-evolution of three main components and their respective actors: permanencies from the original organic structures, interventions during the dictatorship period, and especially interventions after the collapse of dictatorship (Fig. 5.22). From this point of view, this phase, or this collage shows, the survival of the old original principles reincarnated in a delicate new equilibrium that we may regard as a simulacrum of the previous one—that is, similar in shape but still fighting to find a new coherence in its meaning. Continuous redefinition of public/semipublic space through formal/informal temporary short-term arrangements and continuous negotiations became, under the new conditions of limited public control and weak community structures, fragile equilibriums that sometimes even generated social conflicts.

During the past 30 years, attempts to create a new redevelopment culture based on land consolidation formulas and the redistribution of values, which implies a preliminarily approved project, failed. They could not defeat the hidden frame of arrangements and redevelopments of space based on individual informal negotiations. To date, there is not even one success story based on the new redevelopment culture, although governments reacted by enforcing even more legislation and in some cases

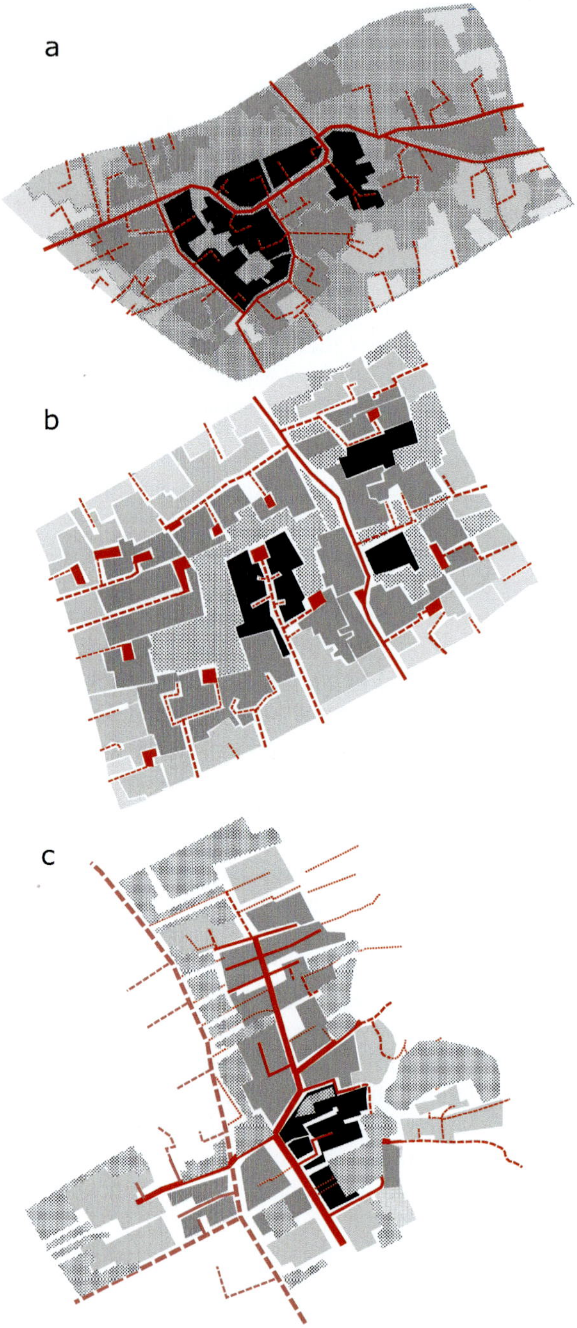

Fig. 5.21 Synthetic diagrams. The three diagrams illustrate how the right to visual privacy acts as a concealing principle at the neighborhood scale by stimulating repetitive deviations of the generative path. This principle is the DNA at the base of the settling process that generated the new emerged reality, reflected in the morphological aspects. The introvert behavior transforms in deviation of the straight line: refractions and bifurcations. **a**, **b** represent circular wrapping. Concealing at the neighborhood level is guaranteed by wrapping around one or more secret cores with limited access from the main paths. **c** represents stripe development along radials that together with buffer isolation from the main roads guarantee the concealing. These examples illustrate the principle that private space is more important than the public space, which is determined by what remains after the private space is defined and developed inwardly

Fig. 5.22 The three pieces collage reincarnated in a simulacrum of the Historical Organic. In the central part, we can distinguish the permanence of the Historical Organic part wrapped by other layers: the four- to five-floor apartments, built during the dictatorship, and the higher apartments in the background. This picture prepares for a better understanding of the next pattern of Recording Over

by using the police to evict people. During the past five years, there are also cases where residents and police entered into violent conflicts, and social tensions were created over imposed developments of this kind. From this point of view, the logic of imposition is very similar to that used during the dictatorship period, but it takes different forms. This means that we need to identify tactical forms of development that fit and co-evolve with the society model, and determine how to turn this situation into a more conscious urban condition.

5.3 Pattern #2: Recording Over

5.3.1 The Essence of the Pattern

Recording Over may be considered as a wave of spontaneity on the premeditated planned sectors of the city (Fig. 5.23). It is characterized by principles and phenomena that aim to give new meaning and function to those parts of the city emptied or "hollowed" by the geometric frames of the apartment blocks built during the dictatorship, in the period between 1944 and 1990 (Fig. 4.16). The spaces secluded by these blocks,

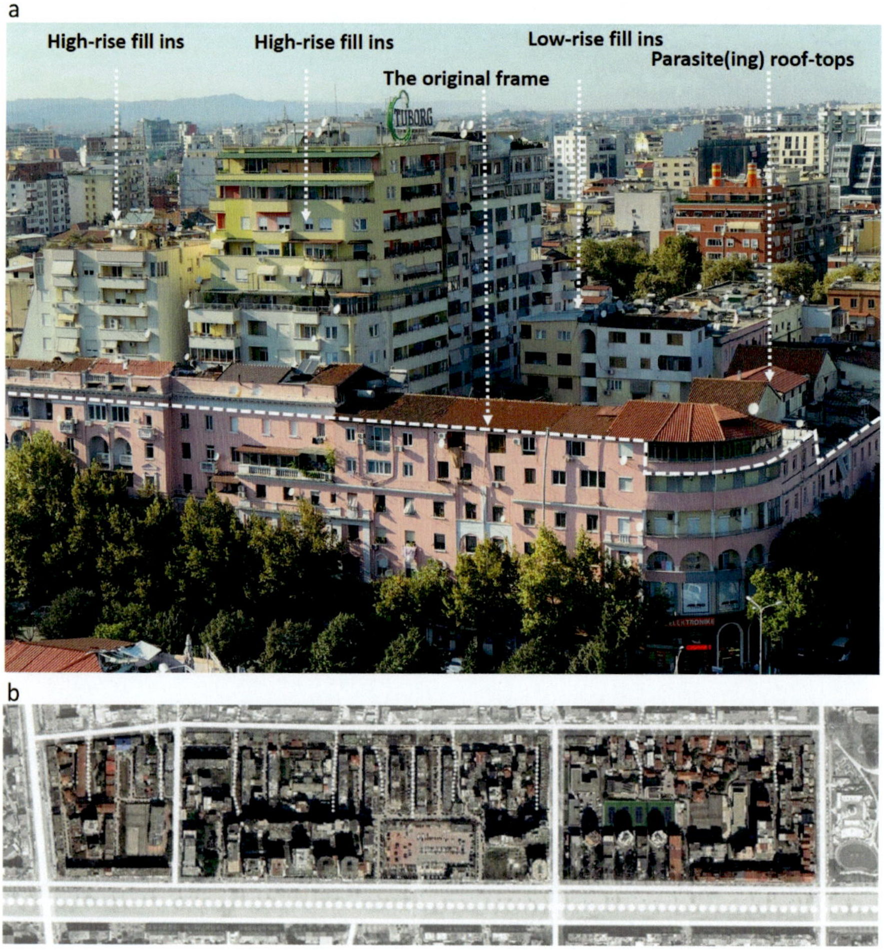

Fig. 5.23 Recording over the representative area; **a** Partial view on *Shallvare*. **b** The urban transect with three sequences: *Shallvare* (built in the late 1950s), *Puna,* and *Gjykata* (built between the late 1960s and the early 1970s). See also Fig. 5.1

perceived as intruders in the city of *mëhallas*, suddenly transformed into arenas of revenge against the public space in the name of the denied past. In this sense, it was a process of deregulating the alien emptiness between frames and internalizing it through self-organized parasite and fill-in structures that organically osmose with the preexisting buildings and the outer-related emptiness within which it occurs, creating as such the involute quality of space. The inward and/or stuck-to-frames behavior remains at the base of this generation process.

This process could not take place outside the presence of predesigned frames. Interestingly, the condition of emptiness turned out to be a powerful catalyzer of those re-appropriative interests expressed in the Recording Over an epidemic process that spread all over the city. From this point of view, Recording Over starts mostly as an informal process, one often driven by corruptive practices, especially when big structures are involved, although a legal process is supposed to be in place.

Despite an appearing social consensus in the socio-political context, the recorded over pattern demonstrates the community's limitations to govern informal arrangements, and the inability to go beyond a small number of families, or limited interest groups. This is evident in its most typical morphological expressions (Figs. 5.24, 5.25 and 5.26), e.g., facades with random extensions/parasitic structures due to different needs or lack of coordination between different families living in an apartment block. As such, the self-organization process in Recorded Over areas is mostly characterized by fragmentary arrangements between individuals and/or small groups. However, at a certain point, especially in the areas where the process of self-governance has been well established, hybrid forms of governance appear. This kind of governance results from interactions of informal arrangements represented by residents (individuals or groups), and formal institutions represented by Municipal government or other spheres of authority (Suhartini and Jones 2019, p. 30). Hybrid governance has been typical in the cases of ground floor extensions used for commercial or business purposes. Common goals are largely shared in this case, and better coordination among residents influenced the municipal government of Tirana to include in the legal planning and design process informal extensions and arrangements. The same practices have been used also in other cases of extensions for housing purposes, if a larger understanding of mutual goals among residents exists.

All these governance types, although manifesting limitations, "overlap, intersect and co-evolve through time" (Suhartini and Jones 2019, p. 224). In these cases, Urban Planning Law or regulations modify to reflect contextual conditions in terms of legalizing informal extensions, and showing a degree of flexibility for "contextual, adaptive and more inclusive governance arrangements" (Suhartini and Jones 2019, p. 232). One such example in Albania was Law N. 9482 dated 0304.2006 on legalization urbanization and integration of illegal constructions, which included the legalization of informal extensions. These are cases when formal and informal not only contribute to the process of legal planning and design, but facilitate future co-evolution between forms of development and a more "nuanced and contextual understanding of governance" (Suhartini and Jones 2019, p. 220).

There are some reasons behind the recording-over deregulation process that started after the collapse of the communist regime at the start of the 1990s. Most of the

Fig. 5.24 Recording Over space-occupiers: fill-in and parasitizing structures in relation to the frames in various positions and stages of recording

Fig. 5.25 Rooftops and other Recording Over elements

Fig. 5.26 Recording Over elements: extensions, staircases, garages, etc. The bottom picture represents the full phenomenology of a parasitized façade

areas where this pattern emerged represent typical examples of an abstract and never completed planning, imposed where narrow, human-scale streets once existed (both physically and mentally). In fact, there were several factors that gave rise to this metamorphic process that involved the ubiquitous emptiness.

First, the changing socio-economic situation and the enactment of a series of laws, among the most important the Law on housing privatization (N. 7652, 23.12.1992) and the Law on the restitution and compensation of assets (including land) to former owners (N. 7698, 15.4.1993), opened new possibilities for the families to individually invest modest finances and incrementally enlarge their tiny apartments. Balconies, rooms, terraces, staircases, etc. extruded from the existing buildings as much as the conditions could allow (Figs. 5.24, 5.25 and 5.26). There were no other realistic opportunities to reach the same objective at that time.

Second, weak public control facilitated the dream of bigger housing spaces and the proliferation of myriad extensions into the Noman's land. Quarrels and conflicts reappeared. Under these conditions, traditions concerning the use of space based on informal arrangements and social consent as the main form of governance were more efficient than regulations and state control, which was heavily corrupted and could not be trusted to guarantee the use of space.

Third, the lack of identity and belonging induced by the emptiness facilitated a fragmentation into parts and a reappropriation process through small-scale adaptations and individual arrangements. This went in tandem with the general sense of abandonment that reigned in both Tirana and Albania during this period. Depopulation as an expression of this phenomenon also involved these neighborhoods. While some people were living, others were arriving mostly from the depressed areas of the country. Residents, who came to these blocks after buying an apartment, were apt to restart and re-form everything from the beginning, including the subdividing and use of the public space. In fact, this process continues to this day in Tirana.

Fourth, in many cases it was an issue of property claims: During the communist regime, many people were expropriated (Law N. 4626, 24.12.1968) without any compensation and their houses were demolished to clear up the terrain for the new apartment blocks. In these cases, people claimed to have rebuilt their houses (or sometimes even high-rise apartment buildings) either in the same place or in those portions of the property that remained free after the blocks were built during the communist regime. In the case of many such claims, the land was restituted (Law N. 7698, 15.4.1993 on the restitution and compensation of assets to former owners), adding even more complexity to the situation. Thus, the organic layer "killed" during the previous period began to be resurrected, piece by piece, sometimes physically separated from the collective apartment blocks and sometimes tacked to them. This layer gradually filled in the public space (considered mostly as emptiness) with new housing or other services.

Fifth, many residents, especially those who lost their titles because of the land nationalization reforms during the communist regime, regarded the occupation of public space through extensions and/or filling-in structures as an opportunity to start up a family business—including commercial ventures, restaurants, cafés, parking lots, storehouses, etc.—and to self-compensate for what they viewed as past injustices.

Sixth, this process of deregulation, or self-reorganization, has to do with the drastic cultural shift (Kostov 2003, pp. 48–50) and the way the space is managed by different cultures. During the dictatorship, the abstract planning blocks created the urban dichotomy, which was represented in the coexistence of two socially exclusive models, one being that of the collective apartment blocks representing the "good" model and the other being that of the remaining substratum of the historic structures made by low-rise traditional houses or villas representing the "evil" to be cured by erasure. After the 1990s, the concept of ownership and the antagonism in managing the space drastically changed, leading to the reconciliation and formal/informal co-evolution of both models. Thus, the self-reorganization of the internal spaces blurred out the dichotomy between planned and organic, or regular and irregular, created by the abstract planning.

Seventh, this kind of attitude toward the public space was at the same time an old tradition in Tirana. As mentioned regarding the Historical Organic pattern, the public space was less important than the private one—that is, it was the result of what was left after the houses were built. This tradition reappeared again, albeit in a more complex situation and under different social conditions. Lack of control, combined with a mentality that did not value the public space, freed residents from impediments on the use of space and activities. People traced secondary networks throughout unused empty spaces and began to set up small, outdoor commercial activities which, initially located along the streets, gradually transformed into modest, sheltered activities before finally crystallizing into new streets and built volumes. Thus, alternative narrow streets generated a new fabric (Fig. 5.27). More than a *cul-de-sac* system, this was a disoriented communication system, one that was often even more complicated than the maze present in the Historical Organic pattern. A parallel city thus emerged, or reemerged from the erased layers, and gradually filled with life and activities. In this sense, there are some similarities in this process with that which led to the previous Historical Organic pattern. The proof that the Recording Over was an attitude toward the public space based on the old traditions is reinforced also by the fact that the same phenomena happened even in cases where the new apartment blocks were built in areas that did not overlap the organic city. The Recording Over layers appeared even in these cases.

Eighth, another important factor for the reorganization of the public space, one triggered by the cultural shift, has to do with the impact of the "public foci on the urban fabric" (Kostov 2003, p. 51). For this, we need to recall the original design concepts of these blocks that, in addition to housing, also contained a social center that either played a mostly propagandistic role or served other administrative functions, including the functions of public libraries, cultural centers, etc. All of that collapsed along with the regime. These kinds of social centers lost their original importance as centers of propaganda and consequently stopped being public foci. Instead, new clusters of commercial services established new polarities that took over the reorganization of fluxes, modeling thus a new course. Often the self-designed appropriative lines flourishing from the new uses obstructed or even blocked the preexisting lines, which were consequently abandoned.

Fig. 5.27 The internal emptied space after being recorded over: from outer related to outer–inner related and closed spaces, including the redefinition of the spatial system and reworked geometry

5.3.2 *The Main Elements of the Recording Over*

The recording over consists of three main elements. The first element is the volumes of the frames, or the apartment blocks which the urban layout followed in an abstract geometric logic in relation to the preexisting context where they were inserted. This simplistic urban design (from 1945 to 1990) considered the city as a mechanic assemblage made of pieces that could be removed at any time. In many cases, these framing walls played the role of urban curtains, especially along the main roads, to mask the historic city considered as a temporary status and subject to demolishing at any time. In fact, frames secluded portions of urban space that gradually became devoid of human relationships and transformed into meaningless spaces (Figs. 4.15 and 4.17), or, in cases where the frames were built *ex-novo*, they hardly contributed to creating a place with human identity. Thus, the Recording Over pattern cannot be seen as separated from the planned interventions. In this sense, it could also be called Organic Deregulation.

The second element is the fill-in structures (Figs. 5.23 and 5.24), mostly represented by free-standing elements that include a wide variety of volumes and functions, such as low-rise housing or high-rise apartment buildings, commercial and leisure activities, garages and parking, open markets, etc. These emerged structures progressively decompose and recompose the sterilized void spaces by gradually giving a new meaning to them. This small- or large-scale invasion process was the opposite of the erasure that took place when the housing blocks, or the frames, were built. Moreover, the newly transformed way of living represented the total opposite of what the absolute planning was meant to create.

The third element is the parasitic structures, which are represented by extensions/extrusions of the inner space of the dwellings in the outer one. They are distinguished by different functions and different positions in relation to the existing buildings. Thus, while the extensions of the ground floors are dedicated to commercial activities, the extensions of the upper floors are mostly dedicated to housing. These extensions can be built at the full height of a building or at separate floors according to the specific needs and agreements of the residents. In relation to their positioning, parasitic structures can be attached to any of the facades, with a preference for the posterior and the lateral facades, or to the rooftop. They can also be extensions or additions of any specific elements such as balconies, stairs, terraces, etc. (Figs. 5.25 and 5.26). These structures added more to the decomposition of the predesigned space and, most importantly, blurred the rigidity of the urban walls/frames built in the previous period. Both parasites and filled-ins were complicit in the occupation of space through informal negotiation or even formal arrangements with the municipality. Therefore, the open-space blocks gradually transformed into involute organic blocks that mostly followed the former organic lines.

At this point, let us explore in greater detail the underlying logic, or the hidden frame, that guided the spontaneous location and the interdependence of the fill-in or parasite structures in the re-appropriative process (using the logic given in Chap. 3, we can consider these as events). For this, we start by identifying the potential Recording Over layers and how the geometry of the different layers was worked and reworked, and co-evolved through formal/informal governance arrangements during this process. In relation to that, there are at least five potential layers (Fig. 5.28) where elements of this pattern can be found or from which they emerge. The first is the former original organic layer (analyzed in Sect. 5.2) represented by sporadic and

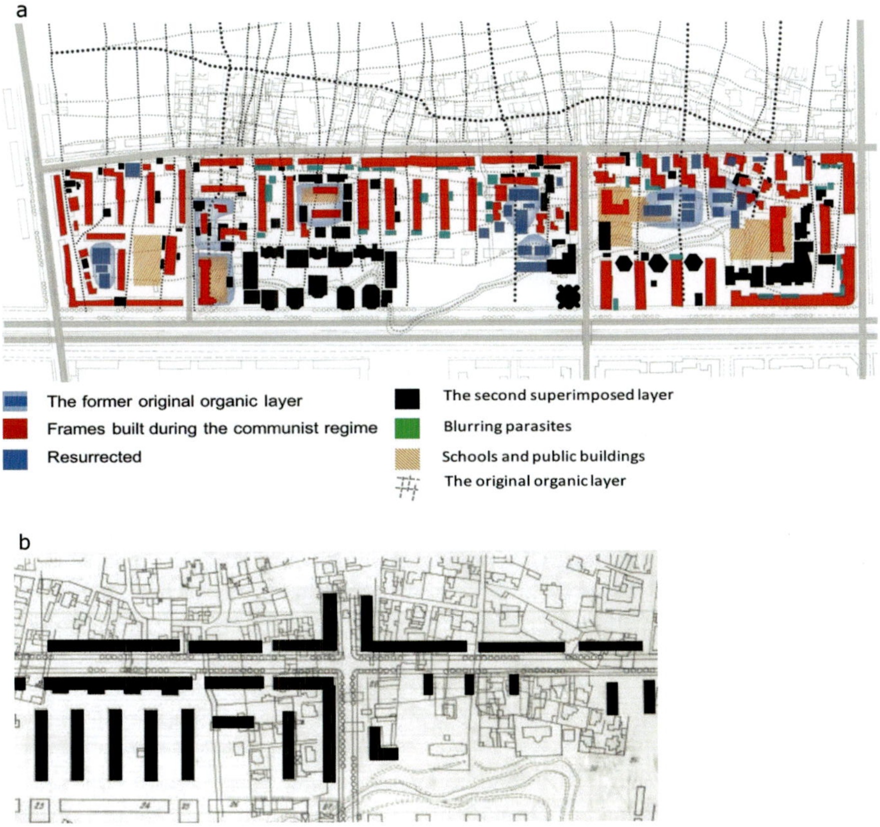

Fig. 5.28 **a** The five potential layers of Recording Over elements. **b** Fragment of the former organic layer with the superimposed frames

fragmented reminiscences from the undemolished historic fabric.[1] Families living there are the main actors negotiating with developers and municipality to develop their plots with high-rise buildings. The second is the superimposed abstract frames of apartment blocks built during the communist regime, or in some cases during the fascist regime, detached from the morphological system and historic traces. The third is the resurrected layer represented by fill-ins built in the space of "reconciliation" between the original organic once erased[2] and the remaining free space still available to be used by former owners (for construction or any other economic activities) who

[1] In very few cases, the historic structure was not totally demolished due to financial restrictions. However, they were considered as temporary structures subject to be demolished at any time. Therefore, in some cases the new buildings (or the frames) were coexisting with the historic structures from the time they were built.

[2] Here I refer to that part of the historic structure that was physically demolished to leave room for the apartment blocks or for the public space but that was not built until the former owners cold re-appropriate.

are the main actors in this case. Fourth is the second superimposed layer represented by fill-ins (mostly high-rise apartment blocks built after the 1990s) that mushroomed in intimate relation with the first superimposed layer (i.e., frames, considered as the only reference structure) or with no relation to anything previous but in areas originally designated for parks or other open-air activities. Developers, former land owners, and municipality are the main actors in this case. Fifth is the blurring layer, represented by parasite structures attached to the existing frames that decomposed their straight lines. In this case, the owners of the apartments[3] (who had nothing to do with the original ownership) benefited from the situation and are the main actors. This condition often translated into social tensions between the former and the newer owners. The government was not able to manage and balance these unknown social dynamics and support the creation of new equilibria. Importantly, these layers need not all be found simultaneously within the same area.

This metamorphic process reflects on one side both changes in the socioeconomic situation and weak public control and on the other side a social harm emerging from the past—that is, denied property rights during the communist regime. This kind of social complex evolved almost into a mass psychosis and motivated the revenge on everything that was considered public. It was akin to the emergence of massive frustration for reestablishing social justice. After years of dictatorship, this was considered a legitimate attitude. Interestingly, despite the efforts of the absolutist planning system to break with the past and propaganda about the evils that private property could generate, in one way or another, there was a large social consent for these informal and formal developments, which could also be considered a form of "therapy" for social relieve. Framing the city was just an unfinished process, or even a regressive process, when it was imposed over the organic structures. In this sense, the Recording Over was an evolutionary process from an immature form and stage of development to one that was more complex, and even consolidated.

5.3.3 The Representative Areas

To understand the Recording Over pattern, we first need to analyze and understand the frames and the conditions under which they were inserted in relation to the preexisting substratum. In this regard, the selected area represents three typical consecutive boxes, or frames, built between the end of the 1950s and the beginning of the 1970s that constitute a longitudinal urban strip that extends from the city center to the first ring (Figs. 5.29 and 5.30).

Starting from the center, within the confines of block 1, *"Shallvare,"* there are two typical structures. First is the so-called *"Italian palazzine,"* a residential block built in

---

[3] The housing reform that started from 1993 (Law N. 7562, 23.12.1992) privatized all the apartments built during the period of the communist regime. Despite the Law N. 7698, 15.4.1993 on the restitution and compensation of assets to former owners of the land where these blocks were built, they were never fully compensated for the land and the buildings that the communist government violently seized (Law N. 4626, 24.12.1968).

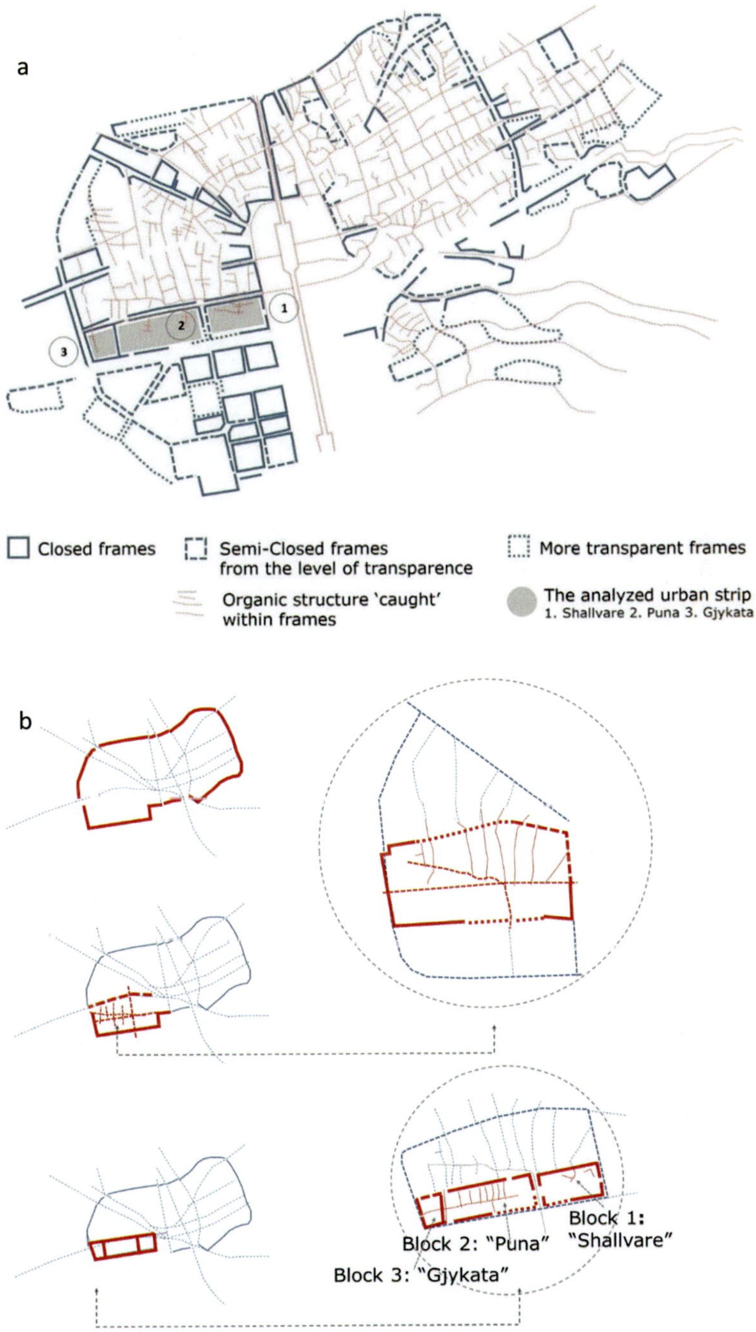

Fig. 5.29 **a** The analyzed urban strip in relation to the main Recording Over frames of the city and the Historical Organic areas; At the city scale these frames played the role of urban "curtains" to mask the historic city. **b** The additional boxes from city to the analyzed urban stripe; The entire city, was considered as something to be "closed" by frames or within "additional boxes" (i.e., one inside the other)

Fig. 5.30 **a** Axonometric view of the analyzed urban stripe as designed by the time of construction *Source* the NTA. **b** One of the closing angles of the analyzed stripe at the start of the 1970s *Source* the NTA.**c** Existing situation of the analyzed urban stripe; See also Fig. 4.25 showing these areas in the map of Tirana. *Source* Google Earth

the late 1930s during the Italian fascist occupation in an area that was free at that time. This block consists of four-story buildings located in parallel with each other within a fenced compound. It embodies the characteristics of the Italian architecture of that period. The second typical structure is "*Shallvaret,*" a residential complex built in the late 1950s under the strong influence of Russian socio-realist architecture. This complex, built in an almost free area, was the largest block of collective apartments in Tirana and one of the first examples that integrated social and commercial services within the complex.

The two other blocks, *"Puna"* and *"Gjykata,"* are typical examples of Albanian socio-realist architecture and urbanism with regard to residential blocks. Their construction, in a partially built area (Fig. 5.28), began in the late 1960s and continued until the early 1970s. These blocks consist of four-to-five-story apartment buildings located at ample distance from one another and oriented according to the eliothermic axes in order to permit light, fresh air, and green areas to penetrate in between. Community services and educational institutions lie at the center of the organization of these blocks. In fact, the figure-ground urban layout (but not the architecture) follows a similar logic with the initial stages of modernist urban design. These three blocks contain the full variety of the phenomena that characterize the Recording Over pattern (Figs. 5.31 and 5.32).

At the city level, the longitudinal stripe was located in the interstitial area between Lana River and the preexisting organic city, which at the time maintained a relative

Fig. 5.31 Phenomena characterizing the Recording Over of the analyzed area; **a** The designed frames; **b** The Recorded Over layer; **c** Recorded Over spaces and frames (rooftop elements are not shown)

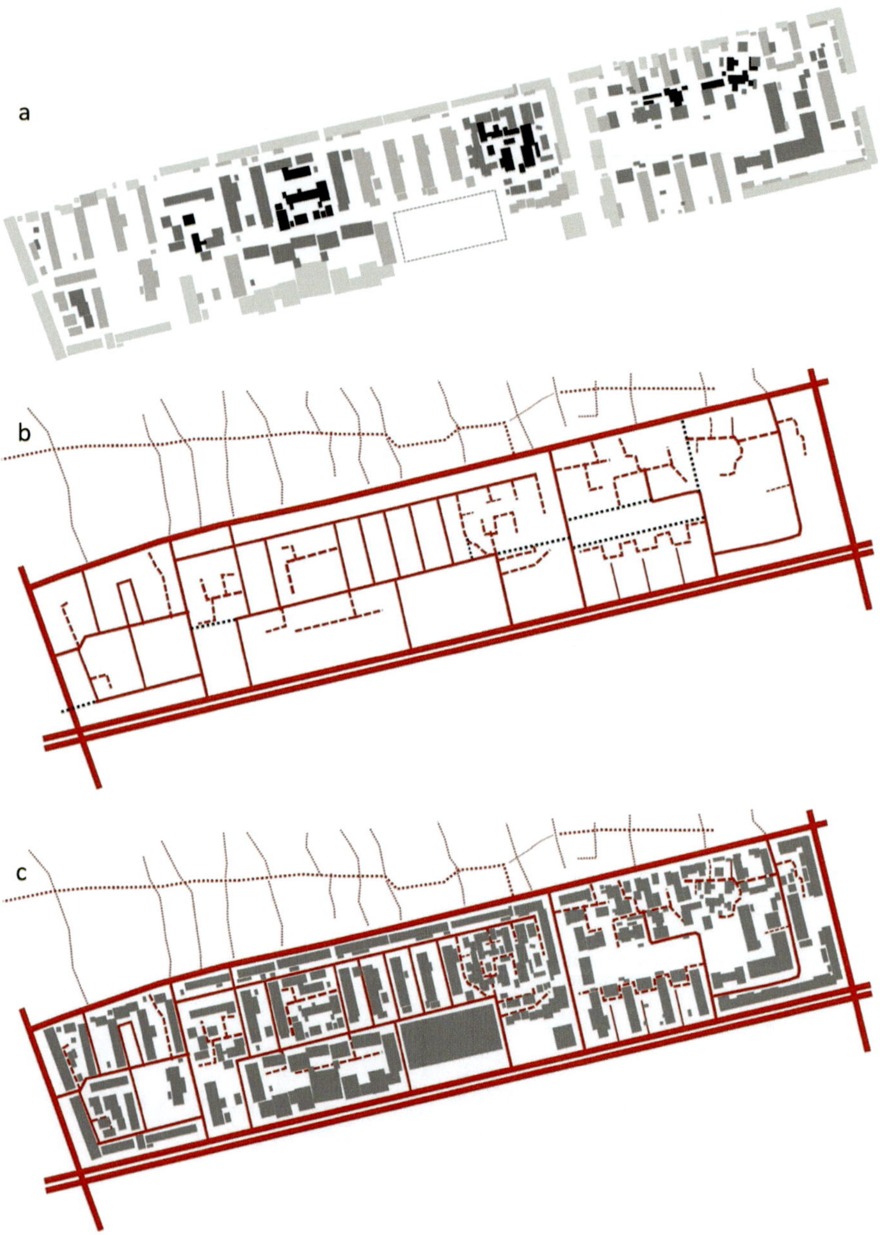

Fig. 5.32 From outer related to outer–inner related; **a** Redefinition and coevolution of the entire spatial system; **b** Reworked geometry of communication; **c** Recorded Over result (see also Fig. 5.27)

distance from the river due to the risk of flooding. In fact, part of the complex's terrain was the previous riverbed, which remained undeveloped until the early 1990s.

The analyzed urban strip was bordered by a road system that was not part of the historic organic structure of Tirana. These circumstances supported an urban design that could be easily detached from the morphologic structure of the organic city. To implement the project, some parts of the organic structures were demolished at the time, especially in areas where the new buildings would be built (Fig. 5.28). Instead, in some other areas supposed to be either open or recreational areas, they remained indifferently unattached and overlaid, wrapped by the newly built structures as a witness to a world that had to disappear. Even in this case, the lack of a binary relationship between the adjacent organic city and the frames was the premise for what happened after the collapse of the regime—that is, a recording over by going back to the traces of the old properties, or even to a dreamed property they never had (Fig. 5.33).

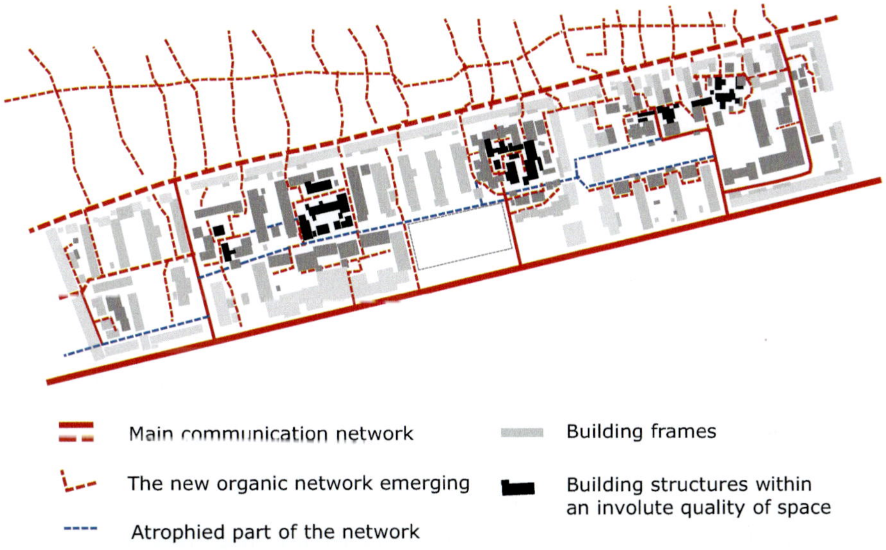

Fig. 5.33 Synthetic diagram: Recording Over emergent structure evolving in a back to organic city scenario. The re-appropriative revenge generates an inward deregulation process and the involute quality of space in a completed status (marked with darker tones) or in progress. Here we can see the continuous wrapping through parasites and fill-ins from within (or from without) to entirely or quasi-entirely interiorize the emptiness; the new meaning of the space within the frames preceding/incubating this process; the evolution from outer related to the involute quality of space through informal/formal forms of governance; the gradual process of re-composition or decomposition of the public spaces mostly following former organic traces, property divisions, or any other pragmatic reasons; and the gradual but steady influence of the Recording Over process on the communication network (deforming, blocking, or atrophying parts of the network within frames) and in the predesigned urban form

5.4 Pattern #3: New Organic

5.4.1 The Two Species of New Organic

As explained, this pattern developed after the decay of the communist regime in 1991, when the phenomenon of rapid and uncontrolled urbanization involved the entire country. For the first time, the compact urban form of Tirana exploded. As such, this event highlighted the passage from the traditional compact city to the configuration of the future metropolitan region.

The New Organic pattern represents the typical characteristics of what normally is classified under the category of informal settlements (Fig. 5.34). The settling modalities combine social improvisations depending on the geographic provenance of the settlers (their anthropological and psychological imprints) with the topographic or land subdivision characteristics of the areas. As such, it shows the lack of predesigned guiding instruments and the vacuum of a central authority. One of the factors that most influenced the characteristics of this pattern is the condition of nearness to the poles of attraction or the main center (Fig. 5.35), and thus further descriptions and analyses represent this condition, which influenced also the essential quality of space. In this regard, the nearer to a center the area is, the denser and the more consolidated it is, and solids dominate the figure-ground contrast. The opposite is observed in the areas located in the periphery.

The need to be isolated was the main motive to select the specific locations and enact the settling modalities, where isolation was desired in order to guarantee a degree of freedom and autonomy from authorities and from other parts of the city that were prejudiced toward or otherwise did not accept them. This psychological condition developed in two different settling patterns at the smaller scale: compressing the space or inflating the space depending on the condition of nearness to the center or to the periphery. Based on this condition, there is a "clashing" or "floating" of elements that generate the corresponding structuring of space. In addition, as will be explained, what needed to be protected from the authorities was not just isolation but also the generation of a critical mass.

I already provided some indication of the dynamics, magnitude, and reasons behind this process in Chap. 4, and I also tried to place this issue in the broader context of natural cities in Chap. 2 by highlighting its organic and emergent qualities, rather than its legal aspects. In this regard, there are some similarities in terms of formal characteristics with the Historical Organic pattern. However, as we will see, there are differences in the motives behind this pattern and also in the anthropological factors that triggered these motives. Following the path of the organic logic, I see this phenomenon as a self-organized coordination at the smallest scale of individual arrangements. The underlying aim that triggered this process was the need of the people to supply housing systems for themselves: their poor economic conditions and the impossibility of finding a niche for themselves in a legal housing market that showed a total lack of public support. The government structures failed to play both, the role as provider and especially the role as an enabler for social housing.

Fig. 5.34 Typical new organic developments near the poles of attraction: Tirana River area in the northern part of Tirana, 2020; (Kristo)

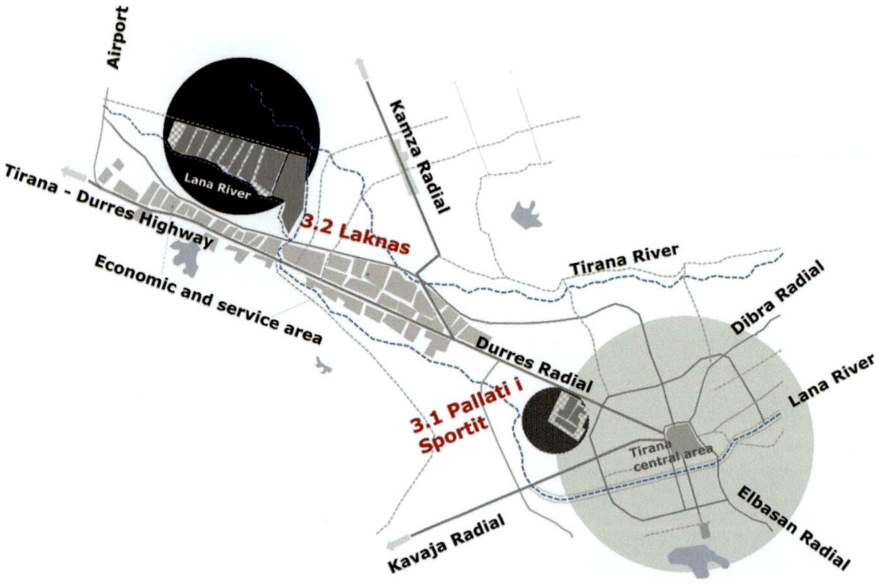

Fig. 5.35 The two representative areas for each of the species of this pattern in different conditions of nearness to the main center. See also Fig. 5.1

This process is still under development in many parts of Tirana. Three decades after the emergence of this phenomenon, we can distinguish areas with different qualities and at different stages of consolidation. Time maturation and the economic incomes of the inhabitants have played an important role in these differences, which are visible also in other elements such as various densities, the degree of fragmentation and distribution of accidental services within the areas, and so forth. Some of these areas are no longer poor and have become more attractive. Interestingly, this organic process where people do or don't do things based on the perceived consent of authorities and/or other members of the community, completely reversed the steps in a so-called normal planning process. Under these conditions, there was nothing left for "planning" and design but to allow this process to continue after failing to anticipate or to prevent it. In this regard, when planning reflects and learns from complexity of the real life, as Suhartini and Jones argue, "governance arrangements increasingly evolve as an organic assemblage of social, political and economic decisions …" (2019, p. 229) to achieve the provision of housing and basic services. Thus, the fictitious claims of central and local government about absolute formal governance in these areas, co-evolved in "nuanced and contextual" urban governance that emerges from interrelationships with informal settlers and other stakeholders (Suhartini and Jones 2019, p. 14, 220).

From another point of view, this illustrates the ability of people to adapt through adjustments in an incremental process. For the same reason, there was no way to prevent the situation from being manipulated in many cases by land speculators. The land was bought through a legal or illegal transaction, or in some cases through a "concession" given by people who were not the real owners but rather land speculators, false realtors, and the like. This created a long-term problem and insecurity not only for people but also for urbanization programs in the area. However, insecurity did not impede people from investing individually in their houses. This kind of socioeconomic and psychological condition creates the informal, formal, and hybrid forms of governance (Suhartini and Jones 2019, pp. 29, 211) interplayed between self-organized individual residents or groups, heads of the group/tribe, land owners (real or false), and other formal institutions such as Municipal government, or international organizations and local NGOs. As mentioned, these modes of governance "intersect and co-evolve." reaching certain levels of "equilibrium" which is fluid and depends on how stakeholders utilize and share their contributions, negotiate or dispute during the development process to provide housing and other services (Suhartini and Jones 2019, pp. 212, 213).

It is important to mention here that people in these areas live also an intense social transformation process. Here we can recall the role of the city as "a machine" that "detaches people from their traditions" and which, in helping to reduce social control, creates more liberty to be different (De Cauter in Architects 2002, p. 11). In fact, this was another important underlying factor for why people went to the cities. Interestingly, this is the opposite of the role that Historical Organic, as pattern, plays today in the city as an event for traditionalization.

5.4.2 Compressed New Organic

The areas belonging to this group are located in a kind of transitory suburban belt, where the urban mass starts losing the quality of compactness but is not fully vaporized. Nourished by rural migrants, the appearance of some basic shelters with recycled materials that mushroomed in proximity to the dismissed industrial plants signaled the beginning of this phenomenon (the top image in Fig. 5.36). The phenomenon rapidly intensified, especially along the main roads, resulting in real settlements getting bigger and denser (the bottom image in Fig. 5.36). Due to that process, and because the city itself was already growing, in many cases these areas were at the center of development interests. As such, they became more attractive for other developments, but at the same time, these factors increased the risks for eviction of the newly arrived residents.

The rise of informal settlements (Fig. 5.37) to meet the need for housing on the one hand and the risk of eviction on the other created the context for the emergence of the phenomena that shaped the quality of "collapsed in itself" in these areas. It is a well-known fact in the organic city generation processes that people group and settle together to create a critical mass for common purposes. This logic, which is one of the reasons why cities are created, is important in this specific case, where people are grouped to defend their housing interests. Metaphorically speaking, they need to build their protective "fortress." Thus, a critical mass and a density turn out to

Fig. 5.36 Bathore area in two stages: 1994 and 2007 (Driscoll)

be crucial for compacting people around a common problem under the condition of the illegal occupation of land. This individual-group (or individual-mass) behavior, which is a kind of camouflage for their non-conformity with the law, aims to increase the number of houses and people up to a critical mass that statistically could change the perception about what is legal and illegal or what is right and what is wrong. In this sense, this was a densification ("fortress sization") process to auto-protect and guarantee their existence under a non-legal status—that is, it was a stratagem to avoid immediate eviction and to enable a metamorphosis from a simple shelter to a real house.

In fact, at the beginning of 2000, the mass of the informal settlements in Albania, Tirana in particular, increased to such levels as to raise serious questions about whether people were violating the law or the law was something that was violating their rights. Informality became almost a common practice to solve the housing

Fig. 5.37 The current situation of the analyzed area (*source* Google Earth)

problem. From this point of view, the amassing strategy worked! The housing issue then was seen also under an ethical perspective and was vindicated as a response to a violated social right. Law N. 9482, 03.04.2006, on the legalization and integration of informal settlements, was a reflection of this socio-political context and a positive sign on a more adaptive governance, supposedly willing to move toward a formal–informal co-evolution model. Unfortunately, this issue still remains open and unresolved. The above-mentioned Law, more than an instrument for inclusion is being used and misused for political reasons and electoral promises.

Under this pressure, within a limited time and limited space, the anthropological principles understood as information about the space distorts and dilutes. Whether people were coming from the same country of origin or not, the respect for tradition(s)—specifically, those related to a patriarchal enlarged family model, ownership, etc.—scarcely survived in their original form. Large-scale emigration shocked the sanctity of family ties by gradually alleviating the obligations coming from tradition and creating opportunities to be exposed to other cultures. Something more important than any specific tradition was prevailing in this case—namely, dwelling and its duration in time. This was related above all to the creation of a critical mass to legitimize "guilt" through, if need be, extreme sacrifices (which happened in many cases[4]). Thus, it was this myth remaining beyond the main motive that triggered this amassing, which aimed to reinforce the "fortress" of their survival. As I will show in Chap. 6, the myth of sacrificing in relation to the built environment itself has profound mythological roots. Under this condition, the shaping process is driven by a pragmatic but at the same time sublime need to overcome the surviving issue that in reality was a legal problem as well.

[4] In some cases, residents threatened to sacrifice themselves inside their houses before they were evicted.

Surviving became the new meaning of the space after traditional historic meanings clashed in such a compacted space–time and collapsed in upon themselves. Thus, isolation and amassing was a way to survive. It is from this "pressurized" condition that the distinctive social behavior of compression derives, behavior that urges densification at the individual scale in order to reach a critical mass. The nearness condition generated after this pressure is more of a clash, or forced intimacy, than a relationship based on anthropological principles (Fig. 5.38). This mechanical aggregation in precarious conditions implies a degree of physical adaptability, both of the public and private space as part of the governance arrangements. This is reflected in the need for flexible and affordable building materials. Two such materials predominated: recycled elements from the ex-factories next to which these settlements started and self-made cement blocks produced within the area, often by families themselves (Fig. 5.39). Similar to the clay bricks used in the Historical Organic areas, the manually produced cement blocks were an affordable and available material that could be put to various uses. In this sense, building materials correspond with the precarious characteristics of these areas or the required levels of flexibility and adaptability. Recently, in the more consolidated areas, new building materials have been used that are similar to those used in other parts of the city.

At a glance, New Organic areas present a high degree of monotony with regard to housing typologies. They mostly differ in their approach to external decoration and architectonic elements such as balconies, roofs, chimneys, stairs, doors and windows, etc. As already mentioned, they began from simple self-made shelters and improved incrementally (Fig. 5.36) until naturally reaching a stage where the demand for an aesthetic dimension manifested on the exteriors of the resulting houses. This phenomenon is expressed through an incoherent and idiosyncratic combination of unrelated architectonic fragments and elements, such as the way they shape the roofs (alpine roofs next to Mediterranean flat roofs and open-air spiral staircases) and the windows, create fences, cultivate their gardens, etc. (Figs. 5.40 and 6.17). These features often represent the imprints of diverse origins. Viewed all together, they create an irrational and bizarre image. In many cases, this interference pattern is manifested as a more shared group experience in order to create clearer signs of group identity within the neighborhood or even the rest of the city. This is typical when smaller groups (approximately five to ten houses) manage to settle together. However, there are many different cases depending on the area.

The one- to three-story houses are the most frequently used typology in the New Organic areas. They range from typical rural family houses, or *elbasançe* (the most diffused typology) (Fig. 5.39) to moderately wealthier two- or three-story houses (Fig. 5.38). All this is done not with architects, but with the help of the so-called master builders. They mechanically replicate the same "project," adapting it to people's requirements.

For a potential visitor, the sense of allowance is similar to that of the Historical Organic. One is attracted, invited, pushed away, threatened, etc., by interpreting signals from the environment. However, the stigma of informal creates additional psychological impediments in relation to potential quarrels and claims for unclear properties that perceptually decrease the sense of admissibility. In addition, a visitor

Fig. 5.38 The nearness condition generated by "pressurized" social behavior. The structures shown exemplify amassing or forced intimacy, which is not based on anthropological principles. Pictures were taken 20–25 years after the process began

is under the psychological pressure of being trapped in an intricate continuity that suddenly may become discontinuity. This is reinforced by the confusing and unrelated architectonic images, perceived as insecure signal, but at the same time as a mystery to be discovered (Fig. 5.40). These characteristics contribute to the social segregation of these areas.

Fig. 5.39 The beginning of the process (1993–1995). These structures were constructed from self-made cement blocks produced within the area, or by families, with the help of the "master builders". Photos Credit: the first, second, and fourth pictures from the top Co-plan archive; the third picture from the top Driscoll

Fig. 5.40 Competing of different imprints: unrelated architectonic elements

5.4.3 *The Representative Area*

"*Pallati i Sportit*"

This sample is located in the nearest peripheral belt of Tirana (Figs. 5.35 and 5.37), where the urban and agricultural fringes intersected in mutual interpenetrations with interstitial unused spaces in between. At the beginning of the 1990s, this was still a well-planned sector housing many public facilities, such as the University of Sports, the Civil Engineering Faculty, student dormitories, the palace of sports and some unfinished sports facilities, the food industry, the old airport, etc. (Fig. 5.41). Soon after 1990s, the alimentary factory and parts of the sports facilities, especially the unfinished portions, ceased functioning. The dismissed activities and the unused physical structures transformed into a great potential for recycled materials, something that advantaged the development of the area, which gradually densified.

The development process started along two preexisting generative paths, both former service roads, going in opposite directions to one another but without meeting at any point (Fig. 5.42). As Figs. 5.41, 5.42, and 5.43 show, each of the two main paths ramified or branched in some lower-level dead-end paths in order to feed the unoccupied areas at that time. As such, they grew as two separated sub-units without intersecting, but knitting internally as an intricate organic network, thus densifying from within to create a critical mass (Fig. 5.43).

Under these conditions, the randomly created free spaces manifested more the characteristics of vacant and unclear spaces rather than those of a real public space. This unclarity echoes the old tradition (of the Historical Organic) that regards the use of public space as secondary and prone to private purposes the moment agreements change. In many cases, these kinds of left spaces, instead of amalgamating separated bodies, are perceived as a space of avoidance, or a potential subject of claims and conflicts. As such, the left space plays more the role of a temporary clash absorber emptied of anthropological meaning (Fig. 5.38) ending in nothingness. This is in contrast to the right to visual privacy as the main characteristic of the Historical Organic dead-ends, although it is formally similar. In fact, the left/"public" spaces are a filter of avoidance for not passing certain limits. This is a new meaning of the

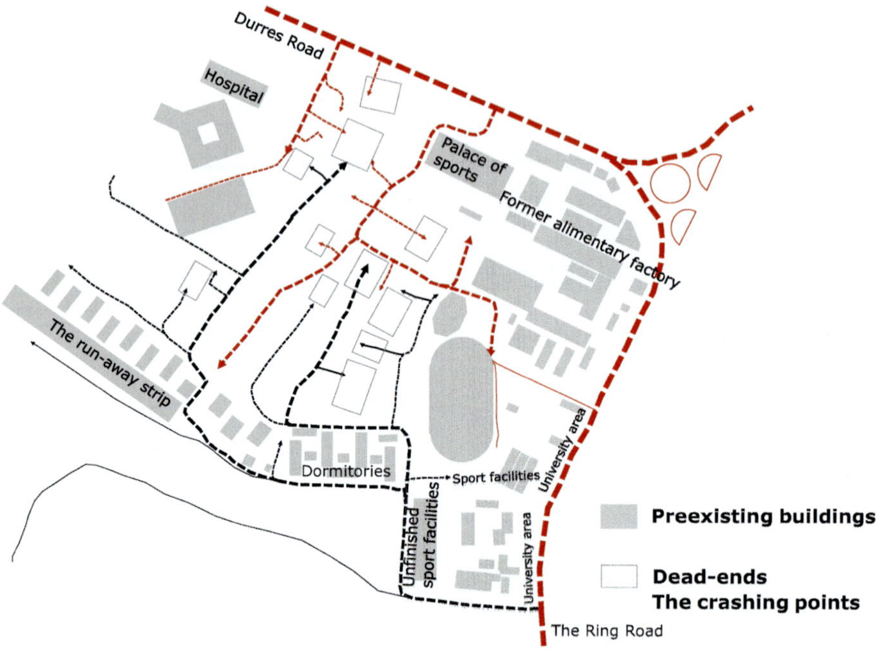

Fig. 5.41 The main preexisting generators within the area. Former service roads transformed into two generative paths going in opposite directions

public space under this specific pattern. However, their peculiarity remains in the status of gradually becoming an identity.

This was one of the first informal areas officially recognized by the government. Urban upgrading projects connected some of the dead-ends in order to create a network (Fig. 5.43) and even created some economic activities as energy sources within the area, supporting a gradual coevolutive process. As I show in the next chapter, these areas emerged through a complex self-organizing process that runs through a series of stages that shift from one temporary equilibrium to another through.

5.4.4 Inflated New Organic

The areas classified under this group are located in the farthest peripheral belt, which is less dense and where the city fabric has almost dissolved. These areas embody characteristics of an open and inflated agricultural city, diluted in the territorial texture (Figs. 4.25 and 5.35) and scattered in the vastness of former agricultural land. The former rural infrastructure (e.g., drainage and stormwater canals and service roads; Fig. 5.44) and current natural features (e.g., riverbeds and green areas alongside

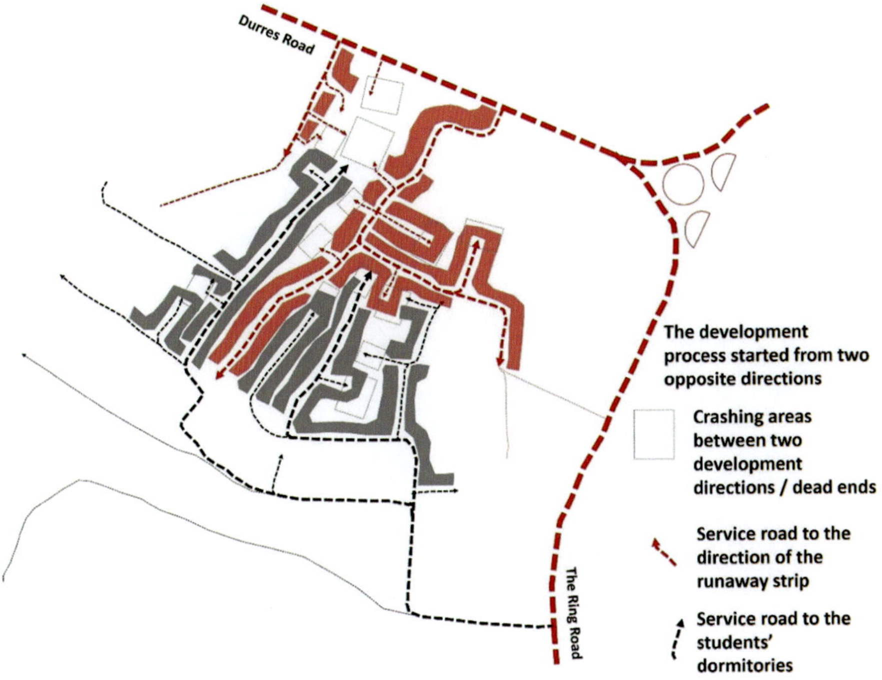

Fig. 5.42 Modality of occupation along the two clashing lines of development

water canals) have guided the location of housing developments in these areas. Other disused structures (e.g., former dairy production farms, military areas, mines, material extraction industries, or industrial plants) have also helped to seed developments, as their infrastructure (e.g., water pipes and electricity) and building materials can be repurposed. This sort of scattered city, like the previous one, is not based on a predesigned scheme, but on the spontaneity of humans and their instinct to adapt through formal and informal arrangements.

The morphology of these areas is a fusion of a large knit rural fabric woven by natural and artificial territorial elements and low-density housing structures. The texture of this fabric combines elements of different scales, which is typical for such areas. For example, the tiny grains of individual houses face large-scale partitions of former agricultural land or natural fragments such as riverbeds or hills, creating an almost dimensional (i.e., out of scale) confusion or a sensation of incompleteness. In these areas, one-to-three story houses are the predominant housing typology, with dictatorship-era collective apartments and other large buildings (e.g., storehouses or factories occasionally found.

Conventional urban categories such as streets, squares, and parks, traditionally used to analyze a city, are not applicable to the Inflated New Organic City. For example, public space consists mostly of access paths, which are embedded in

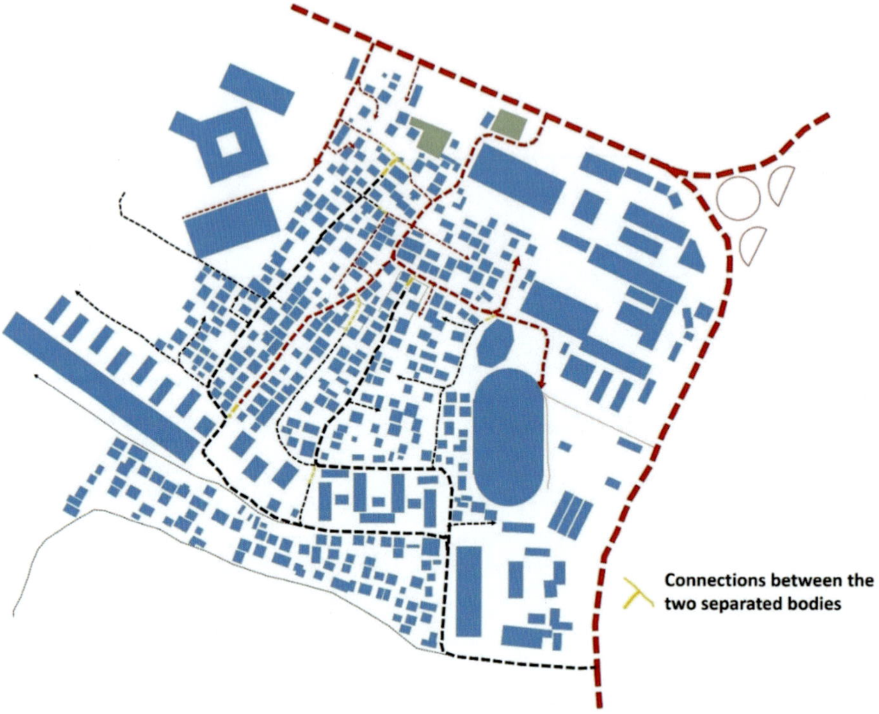

Fig. 5.43 Developments in the area and interventions to connect the communication network

drainage canals, riverbeds, railway tracks, and adjacent areas, as well as in unoccupied agricultural parcels. While this kind of "inflated" city encloses a great deal of free space, which may be perceived as a "luxury," the status of this space is unclear and in many cases manifests characteristics of superfluity. Thus, it is difficult to distinguish between what is public or can be accessed by the public and what is unclearly in a frozen state. While these empty superfluous spaces are potential public spaces, they are not used as such. Further, the fact that these spaces are not used for public activities increases their perceived lack of security. In these areas, the pressure from urbanization was not the same as in areas nearer the center, and the pattern developed undisturbed and under more relaxed conditions.

Most of the newcomers to these areas originated from the northern part of the country or similar regions. The predominant settlement model is much clearer in this case, as the settlers' cultural, social, and historical backgrounds, and thus the anthropological principles based on those values, were relatively homogeneous. Thus, the main characteristics of the settlements of origin (which included northwestern, northeastern, and central northern Albania) consisted of small scattered neighborhoods with isolated dwellings extending across vast areas (Fig. 5.45; Thomo et al. 2004, pp. 27, 75, 97, 190). This traditional dwelling typology involved a solitary house, purposefully (conceived in all its being) not aggregated with similar ones. It

Fig. 5.44 Satellite image of the analyzed area (*source* Google Earth)

could be a single residence or a group of detached houses that sheltered an extended nuclear family. Like small fortresses, the houses were located at a considerable distance from each other as part of an extensive community or mountain village. Within this model, the houses, as focal points in the landscape, were part of an integral (unbroken) territorial system (Fig. 5.45) and grouped around the natural and artificial elements of the wider territory, such as meadows and pastures, groves, creeks and water mills, cultivated land and orchards. Houses could potentially blend into their surrounding natural spaces, but not with other houses, and could include around the houses immense territorial gardens. This isolation had its own historical and anthropological foundations. After the Ottoman occupation, some lowland cities of the northwestern region were destroyed, which forced part of the population to take refuge in mountainous areas, isolating them economically and politically. This revived old forms of tribal organization and provincial units based on common law. Therefore, the isolation of the dwellings in these areas often reflects a desire for safe natural defensive positions rather than any economic reason (Thomo et al. 2004, p. 23).

This was different from the privacy created by deviating from the straight line that motivated the Historical Organic pattern, and it was opposite to the amassing that motivated the Compressed New Organic pattern. As such, isolation for privacy,

Fig. 5.45 The original settlement model: scattered neighborhoods with isolated houses; (the photo at the top Perna)

isolation for amassing (defense), and isolation for loneliness (defense) can be seen as three different settlement models.

These anthropological principles emerged on the periphery of Tirana and drove a settlement model that created isolation through loneliness in the territory (Fig. 5.46).

Fig. 5.46 Replication and application of a similar settlement model on the periphery of Tirana

This quality has been preserved up to the present day for various reasons, including because isolation is embodied as part of the essential logic of this pattern, regardless of the available space; the original plots settled were large enough to offer sufficient space even after further subdivision; the scale of the geometric layout guiding this pattern, such as the preexisting canals, in many cases did not allow a certain vicinity to be overcome; the agricultural/rural culture and lifestyle of these areas, and their related combination of uses (within and outside the private plots) continue to play an important part in the lives of people living there.

The main motive driving dispersion originates from the anthropological factor of isolated loneliness, combined with the need to avoid authorities (defense) and bypass the law, this time through fusion with the territory and "disappearing" within the inflated space in an effort to become unobserved (refuged). As such, the essential quality of this pattern is the dispersed mass of houses and human activities, in a superfluous inflated space. In more concrete terms, space is inflated by combining the condition of superfluity (i.e., an abundance of space) with the actions of dispersion. For its part, dispersion is an escape from "themselves" (families) or the rest of the city and evasion from repressive traditions. Superfluous space, can be inflated by qualifying it through the dispersive model, which in this case is the house perceived as the focus of a larger territory, as already described.

In some cases, we may speak even of a model of rationally inflated space. This occurs when the above-outlined settlement model combines and evolves within the preexisting agricultural infrastructure grid. Even in this case, the house and garden condense in themselves the quality of the settlement model by becoming the "center" of the territory and of the entire "universe," in a regular geometric way. As we will see in the next chapter, in this model, the house is part of an interconnected self-similar territorial system. The composite plot (similar to a small neighborhood), which embodies the idea of a potentially superfluous space, is a complex and fragile universe of relationships based on arrangements, adaptable through actions and reactions in an open system.

5.4.5 Characters of the Inflated New Organic

Character, which can strike us when we visit a city or a neighborhood, becomes an important part of our lived experience in a particular space. In this sense, following the logic of our analyses in regard to the natural city, it is important to demonstrate that the areas merely considered formal/informal, are not only physical spaces but dispone a character. These areas have the full potential to be "places." as Norberg-Schulz says, as an "integral part of existence." and not only as three-dimensional organization of elements constituting the spaces (1980, p. 6). Speaking about "place." Norberg-Schulz says that it is a "totality made up of concrete things" that "determine an environmental character, (1980, p. 6) which is the essence of place" as a qualitative phenomenon (1980, p. 8). According to this author, "character denotes the general atmosphere" as a property of any place which depends on how the boundaries or "space-defining elements" are made (1980, p. 11). Character is also "determined by material and formal constitution of the place" (1980, p. 14).

Thus, the description of characters encountered in the Inflated New Organic areas might help a better understanding of the city or neighborhood, in concrete or qualitative phenomenological terms of the landscape (space-defining elements), especially where the conventional urban categories (e.g., street, square, park) are blurred with the landscape. In this sense, they complement the lack of conventional categories and introduce aspects that may be important for the palace attachment in those areas as "existential foothold" (Norberg-Schulz 1980, p. 5).

These areas embody different characters, here identified as "bucolic," "the day after," and "Alice in Wonderland," (Carroll 1865) as further explained with relation to the sets of images below.

Bucolic

The most pleasant aspects of the Inflated New Organic pattern constitute its bucolic character (Fig. 5.47). This character conveys a sense of abundance and welcoming, from the intimacy of the private gardens to the free open spaces facing the houses. Distant hills, mountain ranges, and necklaces of trees along the riverbeds crown a flat or rolling landscape. High vegetation and abundant fruit trees such as dates, apples, pears, oranges, and plums conceal the houses, which are bordered by vernacular fences. Unpaved former agricultural service roads sheltered by large trees or bordered by reeds create perspective tunnels to the horizon, as if it were a doorframe in a room without walls. Small and mysterious groves, which imply the presence of underground water, are scattered across the vastness. This world, including the creation of space itself, owes its beginning to supposedly free people who, like refugees, sought to escape their own and others' traditions.

The Day After

The character here described as "the day after" embodies the most unpleasant and embarrassing aspects of the Inflated New Organic pattern, while at the same time being quite suggestive (Fig. 5.48, the images on the left). It stimulates a sense of insecurity, unwelcoming, fear, and inaccessibility, and reflects the vacuum of administrative authority. Sculptural cement or necked metallic structures and signs of former industrial buildings dominate this landscape, creating a kind of spontaneous "land art," measurable at the dimension of the open spaces. As such, this character is an absurd combination of derelict industrial or military structures and the current *lasses-faire* social situation of illegal occupation. Isolated young or marginalized people use these abandoned structures in unmonitored ways, including as spaces for leisure and illegal activities. This appropriation of these structures reflects their social frustration and represents a kind of conscious rebellion: a protest against marginalization and a lack of attention from government and social support. Importantly, the potential to transform these "day after" black spots into social attractors and incubators at the city level is substantial.

Fig. 5.47 The "bucolic" character: the most pleasant aspects, creating a sense of abundance

Fig. 5.48 "The day after" (left): the most unpleasant but suggestive aspects of the Inflated New Organic pattern; "Alice in Wonderland" (right): the most unexpected aspects

Alice in Wonderland

The character of "Alice in Wonderland" embodies the most unexpected aspects of the Inflated New Organic pattern (Fig. 5.48, the images on the right). It engenders curiosity and encourages exploration to know "what is there," how to "get in," and to discover "hidden objects" and "oases," visible from a distance but dissipating on approach. This fluctuating clarity, especially at ground level, also produces a sense of anxiety. Objects to discover include small mysterious groves and dense riverside greenery, an unexpected pedestrian bridge crossing the river, experimental agricultural parcels and laboratories, a gated university campus, farmhouses, and vineyards serving as restaurants and wine shops. Other unexpected surprises may be unpleasant or inaccessible, such as former public assets now occupied by illegal activities or improvised housing, well protected by walls like an inaccessible *chateau*.

This kind of reframing is an important vehicle for the transformation of these areas toward a more conscious urban condition. As such, the presented characters may also inspire the normatively designed city.

For an outsider penetrating these areas, the sensations range from mysterious attraction and curiosity to watching beyond the barriers—that is, the barrier of distance or the barriers of dense green agricultural or natural—to a sense of guilt from entering a prohibited space, which reinforces the perception of inaccessibility. This vastness and dilatation of space create a feeling of unachievable secrecy. Walking across these areas, the lack of lateral references means that one feels surveilled, even from above, increasing the challenge of insecurity. These sensations of curiosity and the mystery of the unknown urge further exploration of the "secret core," including the areas alongside rivers and hills, and the remains of agricultural parcels, which look more like abandoned gardens. Nevertheless, there is a lingering frustration at not being able to exhaust the "internality" of the area. The lack of exchange between the central city and this outermost periphery, who are often deemed second-class citizens, further reinforces the perception of inaccessibility, both physically, due to a lack of infrastructure, and more importantly, psychologically.

5.4.6 The Representative Area

"*Laknas*" (Fig. 5.44)

Laknas is located on former agricultural land, once cultivated with various vegetables and fodder. Nearby were two large livestock complexes, part of a state-owned agricultural cooperative. After the collapse of the dictatorship, agricultural land was privatized and distributed to farmers (under Law N. 7501 19.07.1991). This created the first condition for land fragmentation, which triggered an unstoppable land transaction process and further subdivisions. This also opened the way for the construction, without any oversight, of new housing units in areas far from the city center for the first time. This process, combined with domestic migration, created an uncontrolled and chaotic urban situation in Tirana and the excessive use and misuse of agricultural land for construction (Fig. 5.49). The administration was either unwilling or unable to understand these social processes and took no action to help the situation.

People were attracted to settle in the area due to its fertile land, which provided the opportunity to cultivate various crops for livestock. Surveys and reports written by Co-plan (1997) during this period provide evidence that for people from farming backgrounds without the education needed to adapt to the city labor market, this agricultural economy was vital, especially initially. This is one reason why a considerable amount of land in the area remains under cultivation or is used for orcharding or greenhouses. The continued importance of agriculture in the area as an economic activity also explains why space remained "inflated" even after the densification process (Fig. 5.49), in contrast to what happened in the Compressed New Organic areas.

The morphology of the sample area is a combination of low-density housing clusters grouped alongside the former rural infrastructure skeleton, with the accidental but regular free spaces in between used for agricultural purposes. The concept of public space remains vague and superfluous in the area. It resembles an inflated balloon limited by two river branches, within which clusters of houses are dispersed. This morphological character is typical of the Inflated New Organic pattern. A ramified path feeds each of the housing groups, with at least three or four levels of ramification from the main path to the lower-level paths. The area seems regulated according to an orthogonal network, owing to the orientation of the paths and their consecutive branches according to the rationality of the former agricultural infrastructure grid (i.e., irrigation canals and service roads). However, these branches are not interconnected to create a network (Figs. 5.50 and 5.51).

The mushrooming of houses started in proximity to the Lana River and along the agricultural service routes flanked by water canals on both sides, a preference stemming from the need to discharge wastewater. The 90-degree orientation was reflected in the land occupation pattern and positioning of houses within the area that would become the future neighborhood, as well as in the consecutive branching of the service roads feeding the developments. The preexisting gridded configuration and the anthropological principle of isolation through dispersion guaranteed rationalized inflation due to the free interstitials between the houses themselves and between the dispersed groups of houses.

The generation of this structure is based on certain repetitive actions that create the essential quality of the pattern. While the transversal stripes determine housing location, the singular housing elements create the module for a perceptive small mesh knitting (created by the intersection of the housing module lines and the perpendicular lines of the narrower transversal stripes; Figs. 5.51, 5.52, and 5.53). This pixelated grid, subdividing and organizing the transversal stripes, guarantees the essential quality of this pattern. Each unit of these modules could be built upon or used for agricultural or other purposes, creating orthogonal legibility. Each smaller plot could contain the same or similar elements as the original plot, but on a smaller scale, reflecting the model of the neighborhood within a residential plot. The fractal nature of this quality will be discussed in the next chapter.

Fig. 5.49 Settlement models and dynamics in the Inflated New Organic areas (*source* Google Earth)

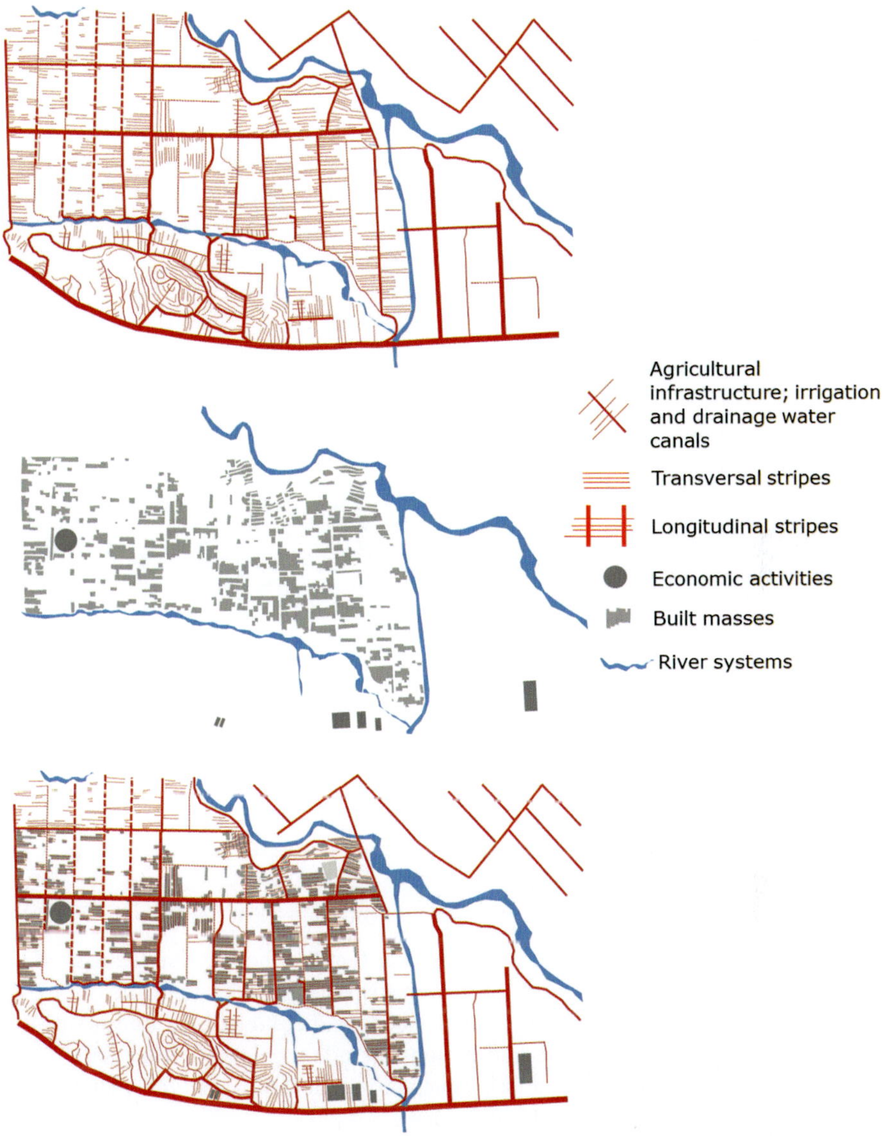

Fig. 5.50 Physical determinants and the development of the area: transversal and longitudinal stripes

The processes of constructing new houses on an original plot (multiplication) and enlarging existing houses by adding more floors (intensification)—which have been taking place in this area in the decades since land reform in 1991 to date—are well "planned" ahead of time. Determining factors include the quantity of the land received by the family during land reform or purchased thereafter and the number of male

Fig. 5.51 Three longitudinal stripes formed between the main water canals and/or service roads; transversal narrower stripes where houses are located; the 90-degree principle because of the preexisting agricultural infrastructure

children expected to remain "within" the nuclear family. The fact that the individual plots correspond to the transversal narrower stripes has given more flexibility to families to use the space according to their needs and conduct land transactions in cases of insufficient space. These transactions include the agricultural reserve each family is supposed to have (Fig. 5.52). Thus, despite densification, this pattern does not create relationships of excessive closeness.

Each plot has two kinds of fences: external fences to designate ownership and internal fences to identify the land designated to each nuclear unit within the family. The former is made of cement blocks, organic materials, or sometimes trees, while the latter is made of sticks or willow. The family plot contains greenhouses and other buildings to support the family business. Toilet services, especially during the initial phases, were established as a separate entity next to the water canals. In some cases, an internal road system nourishes communication of separated houses within the original plot. During important family celebrations (e.g., weddings), the entire plot

Fig. 5.52 Organization of the transversal stripes, which guarantee the essential quality of this pattern; the built and agricultural modules; the condition of distancing or a particular kind of nearness through the different uses of modules (picture on the left Perna)

area is used. This complex coexistence within the neighborhood plot is traditionally overseen by parents or the eldest son.

Figure 5.53 shows the conditions of compression and floating in superfluity as a social behavior. The top image shows how the densified mass of houses in a space that may collapse in on themselves in pursuit of survival—that is, in an effort to extend the life of the settlement against the risk of eviction. Conversely, the bottom image demonstrates how the mass of houses can be dispersed in an inflated space to extend the life of the settlement through unobserved evasion. In the first settlement model, there is a "clashing" or forced intimacy, while in the second, "infinitesimal" subdivisions create a distant vicinity. These are two particular kinds of nearness, not necessarily related to physical distance but to the meaning attributed to the quality of space and the related arrangements.

Before concluding this chapter, I now highlight some differences between the Historical Organic and New Organic, to clarify their distinctness.

Essential Quality of the Compressed New Organic Versus the Historical Organic

Despite the Compressed New Organic being an internally closed space and having some formal characteristics similar to those of the Historical Organic, the New Organic process does not manifest the visceral quality of the Historical Organic. Instead, the space resulting from the Compressed New Organic process creates a forced intimacy of private elements at the smallest scale, becoming potentially conflictual due to the lack of organically established social consent based on tradition. The visceral space, besides being internally closed and protected through concealing

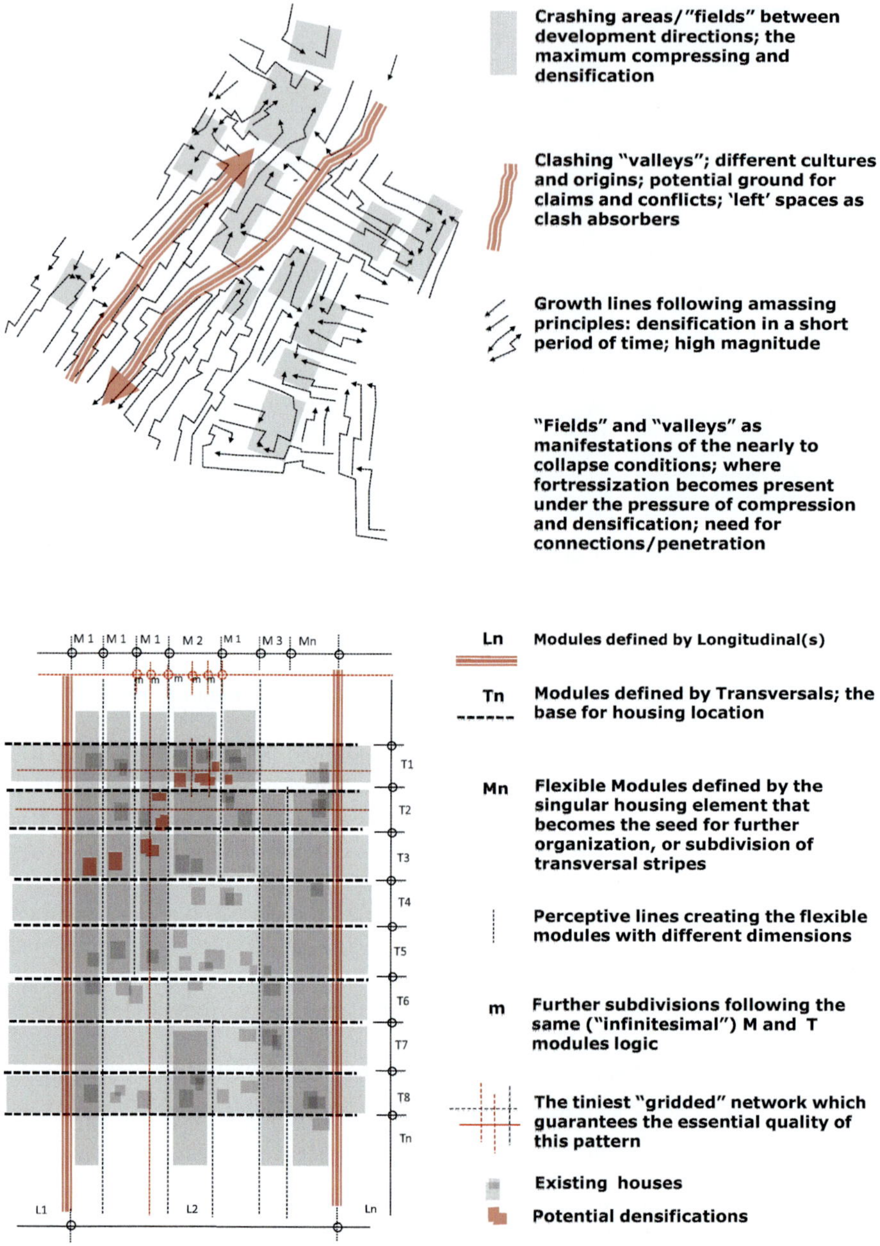

Fig. 5.53 Synthetic diagrams showing two different settlement models. The top figure shows isolation for amassing, or the fortress settlement model. The bottom image shows isolation for loneliness, or the inflated balloon settlement model

itself or deviating from straight lines and open perspectives, is not a conflictual space; social consent based on largely accepted religious principles regulates the relationships. The New Organic pattern does not follow the principle of avoidance and refraction to create the right to privacy; it is mostly a random and forced isolation aggregation to amass. Crucially, it is also a rapid process, making cooperation and the use of a social formula difficult. Thus, these patterns differ both in the meaning they attribute to space and their underpinning anthropological factors.

Essential Quality of the Inflated New Organic Versus the Historical Organic

The Inflated New Organic pattern incorporates an immense amount of superfluous space, guaranteeing the isolation of this pattern and evasion from authorities. In the pattern, dispersion creates a blurring that allows these areas to be lost from sight. Here, there is no need to conceal or deviate; the condition of loneliness in space achieves this end. The characteristics of the Inflated New Organic pattern are transmitted by anthropological factors, which in this case are not completely diluted due to the relative social homogeneity and, critically, reduced interaction with other cultures.

5.5 Conclusions

In this chapter, I presented an overview of Tirana's urban patterns. For each of the three patterns, Historical Organic, Recording Over, and New Organic, I presented the main interacting forces/actors in their historical social context, and highlighted the correspondence between the essential quality of space and the social substance. For this, the representative areas were observed and analyzed in relation to their site-specific emergent quality and formal/informal governance relationships underpinning the self-organized process based on a set of principles and arrangements. This is the first step to illustrate that urban patterns emerge from a complex behavior, a subject that will be elaborated into more detail in the next chapter. Illustrations include maps, synthetic diagrams, and pictures reflecting on the historical or current conditions for each of the patterns, starting from the city to the local scale. The forces that motivate and set in motion the self-organized process for each of the patterns, and their respective essential qualities are summarized below.

The visceral quality of space characterizing the Historical Organic pattern, is generated by a set of principles, actions, and arrangements that aim to close and experience the space internally or from within. This is an implicit requirement of the patriarchal enlarged family in the context of the Ottoman occupation. The right to visual privacy imported by the principles of Islamic city, stimulating repetitive deviations of the generative path, act as the DNA at the base of the settling process. As such it regulated the levels of permissibility based on perceived consent. This pattern represents a specific passage from common to individual: the public space was less important and the result of what was left after continuous redefinitions. This conception of life based on informal arrangements, like a hidden frame, still influences developments, but on a different scale, creating what we might call simulacra of the Historical Organic.

The involute quality of space characterizing the Recording Over pattern, is a set of re-appropriative formal/informal actions and arrangements that aim to internalize the emptiness imposed by apartment blocks built during the dictatorship (between 1944 and 1990). The self-organized parasite and fill-in structures that drastically changed the use of space within these blocks are mainly due to the radical changes in the socio-economic situation and the cultural shift related to it; the weak public control and the inherited mentality that did not value the public space; the lack of identity and sense of belonging to place; and other factors as mentioned in the chapter. Most importantly, Recording Over is a kind of going back to the organic city scenario by mostly following former organic traces and blurring out the dichotomy between geometric frames and historical organic created by the abstract planning.

The two versions of the New Organic Pattern are nourished mainly by rural migration. Their essential quality of space depends on the distance from the poles of attraction in the city. The collapsed in itself quality of space characterizing the Compressed New Organic pattern is generated by a set of principles, actions, and arrangements that aim to group people to settle together and to create a critical mass as protection against eviction. Under a non-legal status, building their protective "fortress" to legitimize "guilt" was the prevailing principle translated in the main building blocks of the space. This psychological environment conditioned the formal/informal relationships and governance arrangements managing the needs for micro adaptations in these areas. The other version of this pattern, the Inflated New Organic, develops in areas located in the farthest periphery and represents a kind of agricultural city scattered in the territorial vastness. Similar to the previous case, this pattern is motivated by the need to avoid authorities and bypass the law. In relation to that, the inflated quality of space is generated by a set of principles and actions that aim to disperse the mass of houses and related activities in the territorial texture based on a model where the house is part of an integral territorial system.

Finally, considering the information presented in Chap. 3 regarding the main theoretical lenses, and the information presented in the previous chapter and in this chapter concerning the site-specific essential qualities of Tirana patterns, we realize that it still remains to be understood, especially regarding the self-organized processes and the way structure and patterns emerge. This is the subject of the next chapter.

References

Architects XDG (2002) After-Sprawl, research for the contemporary city. NAi publishers, Rotterdam; deSingel International Art Centre, Antwerp

Carroll L (1865) Alice's adventures in wonderland. MacMillan, United Kingdom

Co-plan—Center for Habitat Development (1997) Social-economic updating for Breglumasi pilot site, Co-plan documents

Kostov S (2003) The city shaped—urban patterns and meanings through history. Fourth printing, Bulfinch Press AOL Time Warner Book Group, Boston, New York, London
Muka A (2001) Banesa fashatare dhe familja e madhe—Tipi me shtëpi zjarri në qendër. Akademia e Shkencave të Shqipërisë, Instituti i Kultures Popullore, Tirana
Norberg-Schulz C (1980) Genius Loci—towards a phenomenology of architecture. Rizzoli, New York
Shkodra Z (1984) Qyteti Shqiptar gjatë rilindjes kombëtare. Akademia e Shkencave të Shqipërisë, Instituti I Historisë, Tirana
Suhartini N, Jones P (2019) Urban governance and informal settlements—lessons from the City of Jayapura, Indonesia. Springer, The Urban Book Series
Thomo P, Muka A, Zarshati F, Martini GJ (2004) Vendbanime dhe banesa popullore shqiptare. Akademia e Shkencave të Shqipërisë, Insituti i Kulturës Popullore, Toena, Tiranë

Legislation

Law N. 4626 dated 24.12.1968 "On the expropriation and temporary use of property"
Law N. 7501 dated 19.07.1991 "On the land"
Law N. 7652 dated 23.12.1992 "On the privatization of state-owned housing"
Law N. 7698 dated 15.4.1993 "On the restitution and compensation of assets to former owners"
Law N. 9482 dated 03.04.2006 "On legalization, urbanization and integration of illegal constructions"

Chapter 6
Holistic Approach to Tirana Pattern Analysis

Abstract This chapter presents a more holistic analysis of the urban patterns, which are seen as emergent realities generated within a network of interrelationships, or as systems with hierarchical structures that emerge from the dynamics of self-organized macroscopic complex behavior. To adapt these concepts to the processes of human settlement, I introduce the correlation factors of motive, behavior, and modality synthesized in the essential quality of space. Thus, the concepts outlined in Chap. 3, drawn from the quantum, fractal, and complexity approaches, are then applied in site-specific contexts to achieve a deeper understanding of the translation of human behavior into the built environment and to give meaning to emergent urban qualities that would otherwise pass unobserved. In this regard, the chapter tries to clarify the internal world of the self-organized settlement processes in which recognizable patterns expressing the quality of space emerge. The findings of this chapter will then be synthesized to serve as the input for the last part of this book, in which a methodology is elucidated for investigating the emergent realities and their related properties.

Keywords Unbroken web · Correlation factors · System structure · Complex behavior · Self-organization · Emergent qualities

This chapter now presents a more holistic analysis of the urban patterns, which are seen as emergent realities generated within a network of interrelationships, or as systems with hierarchical structures that emerge from the dynamics of self-organized complex behavior. The chapter tries to clarify the internal world of the self-organized settlement processes in which recognizable patterns expressing the quality of space emerges. This holistic approach is consistent with the main purpose of the book concerning new ways of looking at and understanding cities.

For this, I build on the foundation of Chap. 3 where the main theoretical concepts for enlarging the focus of observation along with the justification for applying these concepts to urban analyses were drawn. First, I briefly outline some additional relevant concepts related to wholeness, which will be applied in the pattern analysis. These concepts drawn from the quantum, fractal, and complexity approaches, are then applied in site-specific contexts of the representative areas (shown in Table 5.1) to achieve a deeper understanding of the translation of human behavior into the built

S. Dhamo, *Understanding Emergent Urbanism*, The Urban Book Series,
https://doi.org/10.1007/978-3-030-82731-1_6

environment and to give meaning to emergent urban qualities that would otherwise pass unobserved.

To adapt these concepts to the processes of human settlement, I introduce the correlation factors of motive, behavior, and modality as well as other analytical categories synthesized in the essential quality of space. The findings of this chapter will serve as the input for the last part of this book, in which a methodology is elucidated for the observation and analysis of emergent qualities in cities and their related human properties that serve for a modeling process for better understanding the reality and for use in design.

6.1 Understanding Wholeness Through Relationships

6.1.1 The Unbroken Web of Relationships

In Chap. 3, I outlined the dual nature of matter in the quantum world (i.e., wave–particle duality) and the implications of this for how we conceive the world. For Zohar and Marshal (1994, pp. 54, 62), wholeness is an "unbroken web of overlapping … internal relationships" that generates new emergent realities with new identities. The "unbroken web" is created by the wave aspects that "merge, giving rise to an entirely new system that enfolds the original." This creates the premise for the original and new systems to "evolve together" in a system with its own wave–particle aspects (p. 54). This new system is more than the sum of its parts, as interrelationships are involved in its creation.

In quantum reality, things happen within this unbroken web, which is continuous in space and time. This is because of the arousal effect of the interfering patterns from different space(s) and different time(s) that affect the local temporal environment as no space and time between them exist. As mentioned in Chap. 3, the "holistic" non-local temporal qualities of matter are represented by the wave function that permeates all space and time. This is the quantum correlation effect across space and time, which generates an "unbroken web" of wholeness. Arida (2002 p. 157) called this web the "society-space–time (SST) continuum … an energy field of potential events." In this web, apparently separate "things" are correlated in a "larger whole" across space and time (Zohar and Marshal 1994, pp. 58–60).

Due to the correlation effect that occurs beyond the local temporal reality, the different events in the specific history of a city, anthropological cultures, places, and times interact in the unbroken web of wholeness, generating new emergent realities, including urban patterns. In fact, the human mind produces this corelation, acting as a conductive antenna for the phenomena accumulated in space–time. The human mind stores this information, interprets and transmits it outside the local environment, and even perceives the signals of "wandering" memories within communities. As such, human interactions and their mental imprint, understood as the cognitive space and agents of non-local temporal transfer, play an important role in generating new emergent realities. These realities, owing to the involvement of the interrelationship between the associated waves, are more than physical spatial systems. As humans, we see and transform the world based on our knowledge of it through our mental imprint.

From this holistic perspective, the city is not only a physically static entity to be sterilized by zoning schemes or normative indicators, but also part of a "larger whole," seen through the "unbroken web" of interrelationships (Zohar and Marshall 1994, pp. 58–62), or a "malleable energy field" continuum (Arida 2002, p. 150; see also Chap. 3). As such, the city is a field of information, engraved with and impregnated by invisible local and non-local, temporal, and non-temporal networks of relationships, and material and non-material aspects, to be observed and analyzed. This reconfirms that the specifics of the urban patterns in Tirana (and in any other context) need to be explored within this mental construct: that patterns are emergent qualities that exist beyond the local temporal reality, and that they result from interactions and are not absolute or static entities.

6.1.2 The Relational Order and Dynamic System Structure

In Chap. 3, we saw that form manifests properties of a "system with structure" (Batty and Longley 1994, p. 43) and that the principles of fractal geometry enable one to read a relational order beneath. Thus, taking a fractal approach, wholeness is a relationship between elements in a hierarchical structure, or the recursive behavior of these elements, constituting the internal form that influences the external one.

This relational order, which differs from Euclidean order, explains form as emerging from dynamic interrelationships of elements associated in subassemblies, as the basic components of the system structure. The sets of elements, which define subsystems, may be grouped or assembled "into a distinct hierarchy," corresponding to the statics of the system. Conversely, the relations that describe how the elements "interact and function" explain their behavior or how they correlate (Batty and Longley 1994, p. 43). This is the dynamics of the system and includes forces or any other interfering patterns from inside or outside the system that triggers the internal relational world. These theoretical principles can be considered a guide to exploring form in terms of its statics (structure) and dynamics (behavior), to reveal a different kind of order, which comes closer to giving a holistic understanding of form from a fractal perspective.

Seen in this way, regardless of how irregular a form may initially appear, when viewed as resulting from a process of element interactions, a common pattern of irregularity—which in this case becomes regularity—may be discovered. As shown below, the patterns analyzed in this chapter and their morphological expressions manifest fractal properties; they are complex systems with structure. These systems can be analyzed in terms of their subsystems, which exhibit similarities across scales (or hierarchical structures) and relationships that follow certain behaviors (Batty and Longley 1994, p. 43).

6.1.3 Relationships as Self-organized Behavior

As discussed in Chap. 3, the holistic dimension emerges through relationships generated by a complex behavior with individual and group dynamics at its base. Here, the idea of relationships, already mentioned with regard to the quantum and fractal approaches, is integrated with our urban context. Thus, the correlation effect within the web of wholeness, and the interrelationships of the elements within the hierarchical system structure, will be integrated with the concept of the collective whole, or macroscopic behavior, which arises from the interaction of large numbers of entities starting from the bottom and combines bottom-up and top-down cyclical actions.

To illustrate this in relation to our topic, we first argue that urban patterns emerging from the collective behavior of a large number of individuals to solve their housing problems are expressions of complex systems and adaptive behaviors. The three criteria and properties established by Mitchell (2009, pp. 12–13) help to support this idea. As she argued, complex systems consist first of "large networks of individual components … each … following relatively simple rules with no central control," creating a collective action based on individual arrangements. Second, this happens on the basis of "signaling and information processing" from their environments. As will be seen, signals can also be interpreted according to the quantum paradigm as part of the non-material aspects and the wave effect. Third, complex systems incline toward "adaptation," to increase their "chances of survival or success—through learning" (pp. 12–13). As explained below, survival is a recurrent topic and part of the motivation for the specific patterns observed at the selected sites in Tirana.

The importance of making clear this internal world of relationships from which new emerging qualities arise is evident. Considering these qualities as the field of information to disclose in a holistic way, i.e., as entities manifesting properties not attained by summation, the analytical categories to unpack these qualities should include interrelationships and correlations from where this reality emerges. Hence, these categories should reflect the dual nature of the wave–particle or the non-local temporal properties of quantum particles interpreted by Arrida as the event and the associated event horizon (presented in Chap. 3). From this mental construct, I infer the category of the original event as the initial "point" existing in the holistic dimension of the interrelated time–space–society web, generating the motive as the wave factor associated with it.

In addition, these categories should reflect the structure that emerges from interrelationships of elements behaving in a certain way within the system, and the way they group into hierarchies and repeatable/detectable patterns. From this, I infer the category of behavior that not only links with the categories of the event and associated motive, but most importantly reflect the relational or fractal perspective as presented in Chap. 3 and at the beginning of this chapter. Most importantly, the behavior, or the way elements correlate, is based on mental patterns transported by this motive that create specific geometric configurations at the smaller/individual scale, as explained and shown later in this chapter.

Finally, these categories should reflect how all this happens within the collective whole emerging from the world of internal relationships, and how the individual behaviors translate into shared behavior in an approval/disapproval process through levels of hierarchy. From here, I infer the category of modality that links with the previous ones and as presented further in this chapter, embodies the full cycle of interactions and adaptations starting from the bottom to the feedback loop from the top (as the winning behavior).

To this end, I introduce the correlation factors of motive, behavior, and modality, which are impulses, signals, and actions within the "web of wholeness," or in the "space–time continuum," which act through levels of hierarchy as part of this complex process. These factors activate, organize, control, regulate, randomize, and even decline in disorder the entire generative or degenerative process. They are intimately related to the essential qualities emerging from this process and can be used as the base for the analyses presented in this chapter, and to address a higher level of understanding after what was already described in terms of the preliminary analysis in the previous chapter. Correlation factors of motive, behavior, and modality, amalgamate concepts from quantum, fractal, and complexity theories. These factors are explained below.

Human motive is the wave factor associated with an original event transcending time and space. In general, motives are generated by preconditions deriving from social and historical contexts (past-present) and forge the final aim in relation to specific space creation. It is translated and transmitted as anthropological principles pervading the space–time web (wave) and remains as a property of it, appearing and reappearing, and guaranteeing non-local temporal qualities. Drawing on Kelso et al. (2016, pp 44–45) description of how parts of complex systems interact to produce structure and pattern, motive can be compared to their system "control parameter." In concrete self-organized systems, motive promotes the interaction between parts by triggering individual behaviors driven by this motive. This book refers to both motive and control parameters, with the latter used to describe process logic in general, while motive is associated with human properties. Motive also embodies the human reasons for how a process is triggered and the origin of the event.

Behavior is the individual interfering pattern imported in the local temporal horizon and detectable within this horizon or reality. As mentioned, this intrinsic property of the system is triggered by anthropological principles or the motive (past–present) and is translated into individual arrangements as the horizon of these principles. These horizons represent the establishment of partial equilibriums that compete with each other until a certain threshold is reached (Kelso et al. 2016, pp. 44–45). Thus, individual behavior is a vehicle contributing to partial equilibriums, arising from the continuous competing of individual and group arrangements up to a certain limit, at which point individual behaviors become a social behavior or shared experience. As such, behavior can be considered the dynamics of the internal part of the system structure that acts at the bottom.

Modality is the way the main motive is achieved at the macroscopic level through translating the competing individual behaviors into recursive social patterns (e.g., collective intelligence or swarm behavior). Modality is akin to what Kelso et al. (2016, pp. 44–45) termed an "order parameter," which operates at the macroscopic level of the system and "enslaves" the behavior of individual parts by imposing common logic and translating the motive related to the essential quality of space into a social norm. Modality is how a behavioral pattern is concretized into a social norm based on information exchange from competition at the lower levels of the system to achieve a motive. This recursive action can also be considered feedback from the top of the system,[1] or a kind of "generator," applied to the "initiator" to replace and scale it, creating subassemblies comprising groups of elements repeated in a self-similar or self-affinitive way (Batty and Longley 1994, pp. 47, 61, 62). An example of this can be seen in the repetition of the refraction of street lines and the settling of the housing plots according to a majority-accepted modality.

The essential quality emerging from this complex process that produces structure and pattern is a local temporal reality arising from a motive (i.e., the associated wave factor of an original event, or "control parameter" that pervades the system), the behavior (interfering pattern at the individual level, or dynamic at the bottom that creates partial equilibriums), and a modality (recursive social pattern, or "order parameter" at the top). The essential quality can be considered the finer scale in the local temporal reality, where the idea about the space translates into the space due to this generative cyclical bottom-up process. Once the behavior of the individual parts is "enslaved" (Kelso et al. 2016, pp. 44–45), it tends to translate into a mental pattern or archetype, lying dormant in the human brain to emerge in another time and/or space to be recycled through the web of wholeness. As Portugali (2000, p. 58) argued in "the paradigm of pattern recognition," cities, or parts of them, are self-organized systems, both physically and cognitively. In the cognitive system of a person, the process is analog. The order parameters of several configurations "are formed by means of associative memory" (p. 58). They "enter into competition," and when one of them "wins," it "enslaves the cognitive system" (p. 58). This mechanism is crucial for the way these categories (control parameter and order parameter) have been translated into visual codes as shown further in the chapter.

To further define this kind of complex systems, according to Holland (2014, p. 35), they are Complex Adaptive Systems (CAS), in which the elements or agents "learn" and "adapt" through interacting with other agents. Holland explains how "signals" and "actions" (pp. 57, 58) are confirmed or disconfirmed passing through levels of hierarchical organization of CAS. Their boundaries are like semi-permeable membranes that guarantee the passage of those signals that syntonize with the motive (pp. 64, 65). As shown in Sect. 6.4, this cyclical self-organized process within CAS acts like a feedback loop for the selection of the successful standards (p. 71).

[1] The concepts of the bottom and top of the system come from the fractal understanding of the system structure, explained in Sect. 6.3.

Pattern analyses presented further in this chapter synthesize the combination of correlation factors (motive, behavior, and modality which carry aspects from quantum and fractal theory), Portugali's (2000, pp. 57, 58) "paradigm of pattern formation" and "paradigm of pattern recognition," Kelso's et al. (2016, pp. 44–46) "synergetics" and "inter-representation networks," and Holland's (2014) CAS theory. This conception of the self-organized settling processes is coherent and explains the internal mechanisms triggering the range of the governance relationships and formal/informal arrangements (Suhartini and Jones 2019) presented in the previous chapter. Using the combination of these theories in the analytical process we may address a higher level of understanding of the urban patterns from a holistic perspective.

The high degree of incoherence in Tirana (both at the urban and architectural scales) is due to the interference patterns of human users, who come from different areas of the country, have different cultural backgrounds, and whose memory stores are used in a continuously competing and adapting interrelationship (Arida 2002, p. 161). Memory as information can include fixed ideas, frustrations, and projects, resulting in a local interpretation. Figure 5.40 from the Collapsed New Organic shows the incoherence of architectonic images due to the competing of individuals and groups with different cultural backgrounds and behaviors, and because of the associated wave factors from different events. The resulting mental patterns and archetypes remain in the competition.

Having presented the main principles and mental constructs for more holistically approaching the essential qualities of patterns, the importance of the notion of relationship, with the additional meanings extracted from the quantum, fractal, and complexity approaches, is apparent. From this foundation, we can now identify the generative process behind each of the patterns treated by this book. Here, the city is not a physical space emptied of significance; it is a complex wholeness generated from actions embedded in the web of human properties, knitted by motives, behaviors, and modalities. Users shape space, while their brains are simultaneously engraved with ideas about space, influencing their behavior in relation to space. In this way, the mindset emanating from quantum wave–particle duality adds further significance, meaning, and depth to observed phenomena.

6.2 Quantum Approach—Patterns as Overlaps Within the Unbroken Web

6.2.1 *Historical Organic as a New Emergent Reality*

Original event: *imaret/kulliye* system
Motive: the right to visual privacy (associated wave effect of the event in the local temporal reality)
Behavior: introvert (interfering pattern in the local temporal reality; at the bottom)
Essential quality of space: visceral (the emergent event horizon/reality); Fig. 6.1

Translating the theoretical concepts presented above into analytical tools begins with envisioning urban patterns and their morphological expressions as not only physical entities but, more holistically, as emerging from systems of interrelationships. Then, considering the essential qualities of patterns, already presented in the previous chapter (visceral, involute, collapsed, inflated) as observable site-specific entities in the local temporal reality, caught within the unbroken web of relationships, the challenge becomes to explore their morphological expressions as horizons arising from the associated wave effect of the original event (i.e., the motive) and the individual behavior and social modality from which these patterns derive. This relationship creates the "correlation effect" due to the absence of intervening space–time (Zohar and Marshall 1994, pp. 58–62).

There is an intimate relationship between the visceral quality of space and the model of life rooted in socio-historical preconditions and traditions, especially the concern for visual privacy. As presented in Chaps. 4 and 5, this tradition directly concerns how public or private spaces flow or interlock. As such, it is important to investigate community connections and networks and the anthropological principles guiding the settlement process, as these underpin the interrelationships that create the unbroken energy field connecting the past with the present and influencing the future. Likewise, they motivate and trigger specific human behaviors according to the same or (self)similar principles. Understanding these immaterial aspects can thus provide insight into the factors influencing public and private life.

The above-defined categories of motive, behavior, and modality are used to explore the holistic interrelationships generating the visceral quality of space. Thus, first, we must understand which incubators forged and transmitted motive, guaranteeing the system of interrelationships within the unbroken web of wholeness. As explained in the previous chapter, the social structures that generated this pattern were based on family ties and affiliations and led to the formation of the first urban clusters (Figs. 4.8, 4.9 and 4.10). The family was the smallest unit, within which the

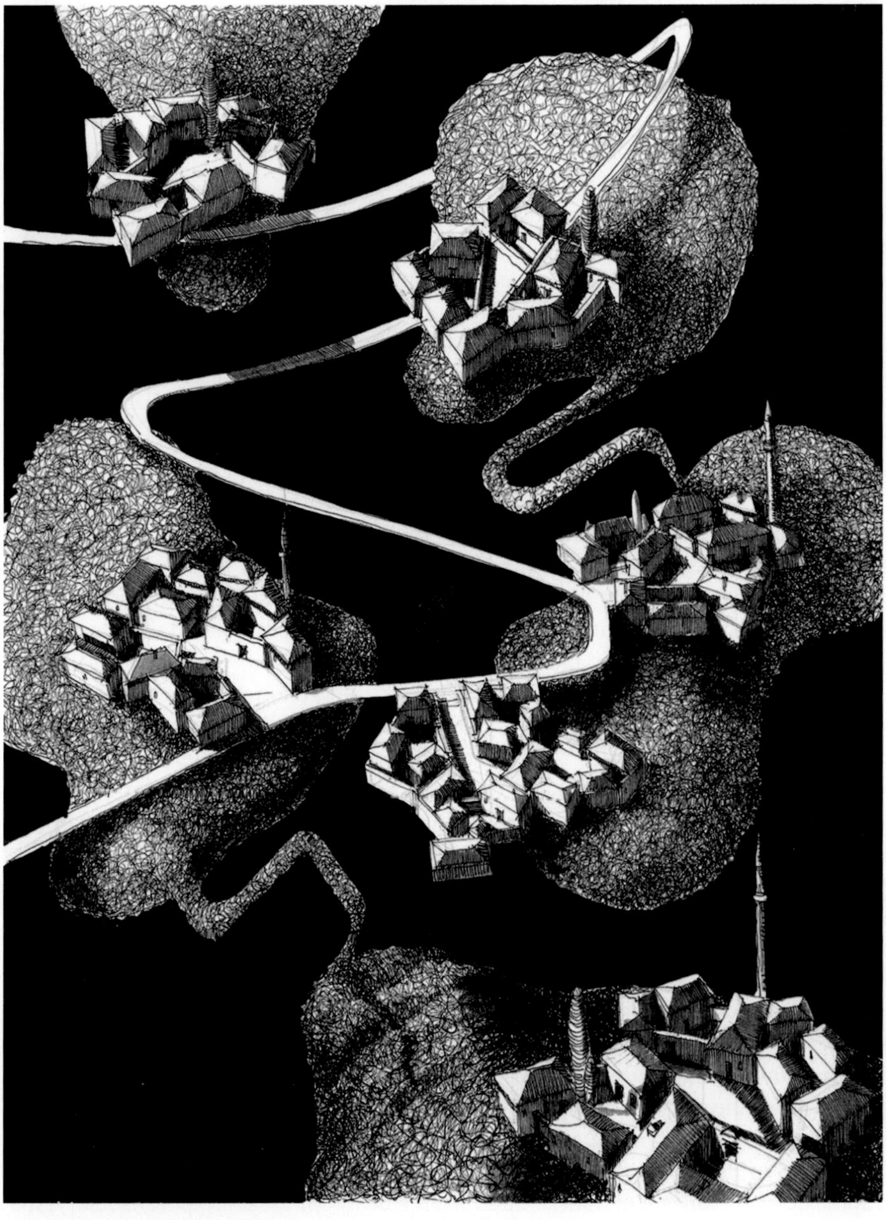

Fig. 6.1 New emergent realities in the web of wholeness; artwork by Endrit Marku, especially designed for this book

rules of tradition were transmitted and reproduced as a means of continuing relationships replicating a model of life that corresponded to the livable space. In Tirana, the traditional *imaret* and *kulliye* systems, which promoted the Ottoman way of life, established a "concern for privacy" (Kostov 2003, p. 63) as an important driving factor during the settlement process, making it instrumental in all life processes and interrelationships, both within and outside the family. This concern is reflected in the avoidance of "visual corridors," starting from a cluster of houses (Figs. 5.2 and 5.3) and extending to the urban vistas at the city scale. This gradually emerged as the main motive behind the material expression of internal and external urban space. In this pattern, the family was the fulcrum that produced and maintained the invisible "filaments" of the unbroken web of wholeness (Fig. 6.1). Through these filaments, families acted as non-local factors by creating the transmission tunnels in space–time that recycled the principles incubating the main motive—the right to visual privacy (Kostov 2003, p. 63)—as an associated wave factor aroused within the unbroken web, which in turn generated the individual arrangements at the bottom (i.e., introvert behavior) and imposed a common logic as a recursive social pattern (i.e., a shared modality of refraction) from the top. This introvert behavior became an interfering pattern in the local temporal reality: aggregation occurred internal to occupied parcels, and great importance was placed on closing visual corridors and filtering contact with the outside world, despite occurrences outside these areas and with no regard for the public interest.

As described in Sect. 6.1.3 concerning the interaction process within a self-organized collective whole, once individual behaviors are transformed into a shared identification, they cyclically reproduce across scales. Thus, once the community living in a neighborhood consented to preferencing closed perspectives and deviation over opening and creating continuity and began imposing this, the modality of avoidance through refraction and bifurcation was cyclically repeated across scales, becoming an intrinsic value, or part of the DNA, of the unbroken web of wholeness. This process and the values that permeate the entire system by repetition of the same principle across scales further reinforced the introvert character (Figs. 4.8, 4.9, and 4.10, images at the bottom) of the house and entire residential cluster. The appearance toward and from the street was not important (Kostov 2003, p. 63); the priority was to conceal. This concealment and avoidance were reflected in the relatively high walls confining plots, reduced building height (one or two floors), the use of dense greenery, and the house's distance from access roads (Fig. 4.2).

The Historical Organic pattern in Tirana was a new emergent reality generated by the overlap of the associated wave effect of local events (mainly physical and topographic determinants) with that emitted from a different local temporal society; that is, the *imaret* and *kulliye* traditions, which embodied a set of anthropological principles imported by religious missionaries into Albanian territory during the Ottoman period. Once initiated in Tirana, the *imaret* and *kulliye* systems became a new local event with its own associated wave effects or horizons (e.g., houses, services, administration, and activities), which in turn generated a motive for urban knitting based on the right to visual privacy. Thus, each stage of development became a frame of

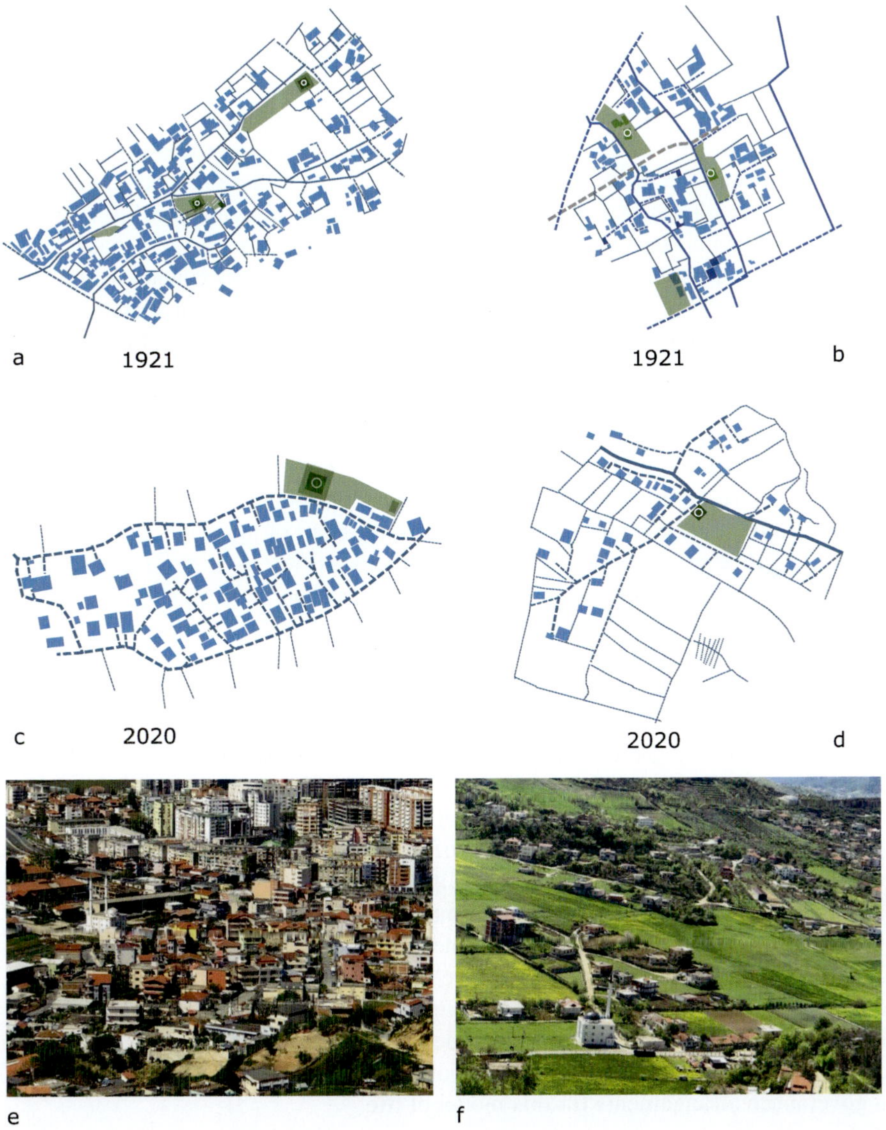

Fig. 6.2 The images at the top **a** and **b** show the *kulliye* system of 1921 as an event; **c–f** event horizons of 2020. The historical organic pattern, which itself is an emergent reality (horizon) aroused by an event existing in another space–time reality (the Ottoman *kulliye*), plays the role of the event transcending time and space and materializes similar principles in 2020. The territory is not a fixed entity, but rather a reality emerging from interrelationships beyond local temporal realities

events influencing further developments through its associated wave factors, creating new horizons or emergent realities based on new or transformed motives (Fig. 6.2).

The Historical Organic pattern, which descends from the *imaret* and *kulliye* systems, remains present in modern-day Tirana, both physically and immaterially, as part of the web of wholeness. This permanence plays a role in the physical events that, through interfering patterns, continue to influence aspects of public and private life in the city, as well as the approach to settling and building processes. Moreover, event horizons are reproduced as physical expressions of these mental frames (i.e., the emergent realities). An example of this is the recent pragmatic infilling within Historical Organic areas as a simulacrum of the visceral quality of space in the Historical Organic pattern. The order beneath these new developments is not only physical but an expression of the mental imprint of the traditional system. This is illustrated in Figs. 6.3 and 6.4, which depict multistory collective apartments choreographed (or perhaps trapped) by interrelationships controlled and ordered by the interfering patterns of the Historical Organic. The geometry of distance is the only controllable instrument for the insertion of high-rise typologies, even though this is negotiated with traditional style and evidently often not respected. Paths containing a great deal of refraction and bifurcation, often making access complicated, feed the high-rise infill developments, which penetrated the heart of the neighborhoods. This is because public space continues to be conceived of as "left over" and less important than private space, a situation stemming from internal aggregation.

We can conclude that the right to visual privacy as the main motive generating the visceral quality of space through an inward aggregation process emerged as an associated wave factor of the *imaret* and *kulliye* systems, the original events existing in the web of wholeness. This motive is triggered in the local temporal reality of Tirana interfering patterns expressed in the introverted behavior of individual arrangements in the settlement process. The resulting interactions and information exchange gave rise to a shared modality at the neighborhood (system) level, which was expressed in the recursive refraction of straight perspectives (streets) during the settlement process. Thus, the visceral quality of space is the horizon of the non-local temporal effects rooted in specific religious culture and anthropological principles (the *imaret* and *kulliye* systems). The principles that forged this motive emerged in a different local temporal reality and were seeded in Tirana through the correlation effect within the unbroken web of wholeness. This motive organically drove a self-regulated urban knitting, i.e., a labyrinthine structure that became the essence of the city form, representing the visceral quality of space—the main requirement embodied in governance arrangements for this model of life.

Repertoires of Motive—Generative Apparatus and the Genes Beyond the Patterns

How does the concern for privacy materialize from a motive, or associated wave, in physical space? Is there a repertoire of mental patterns or archetypes that fulfill the requirements of the motive—as genetic expressions of the desired quality of space? In answering these questions, we see that mental patterns based on previous configuration experiences or associative memories, are initially a device transported by the human brain (i.e., interfering patterns) and serve as the mental building blocks for the configuration of urban space. Due to the generative bottom-up processes driven

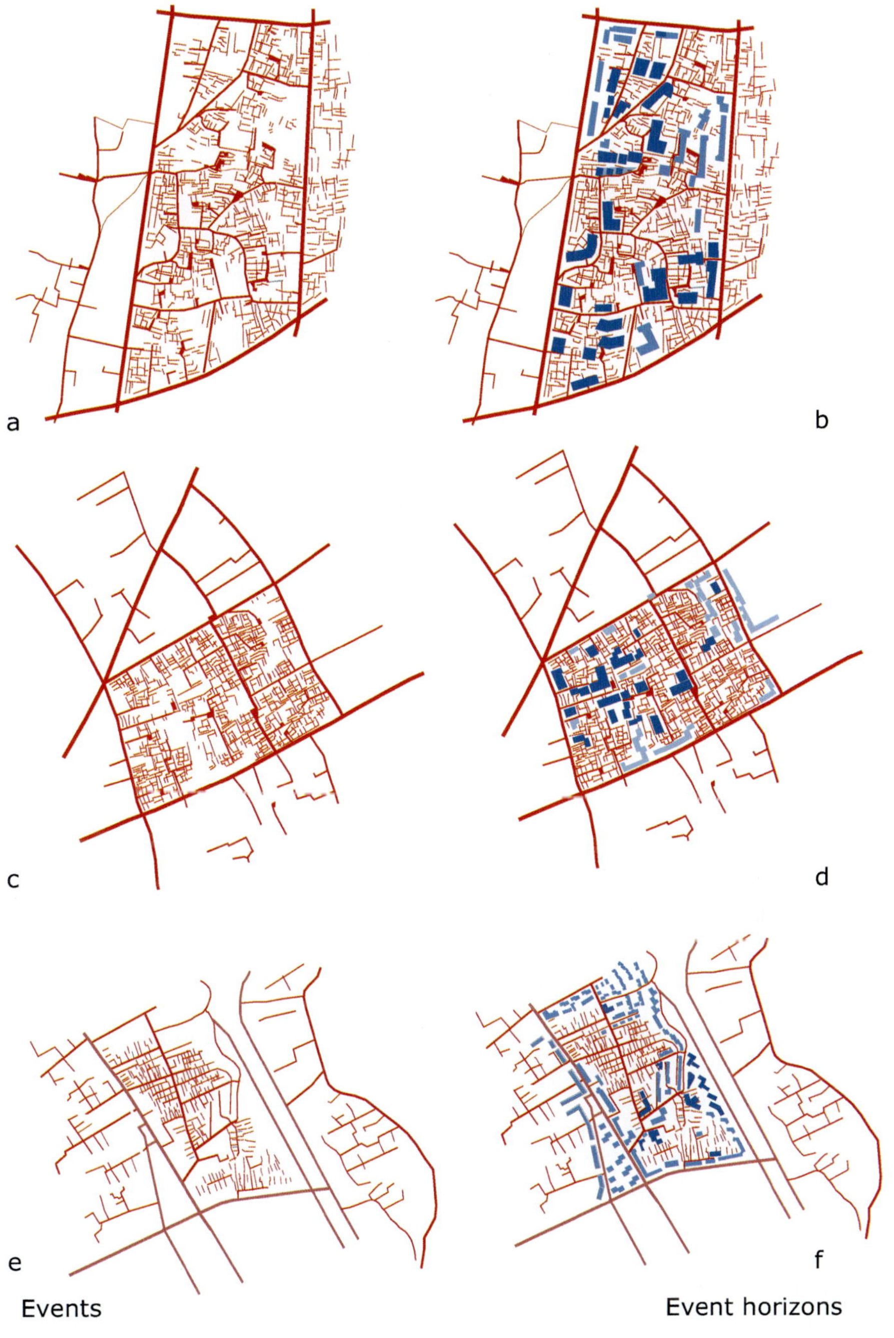

Fig. 6.3 The images on the left **a**, **c**, and **e** show the urban knitting as a hidden frame of events with its own wave–particle aspects generating new emergent realities. The images on the right **b**, **d**, **f** illustrate the present situation of the historical organic as an event horizon or simulacrum of the original historical organic

Fig. 6.4 Multistory apartments trapped in the historical organic maze

by the specific motive, these devices give rise to a series of geometric configurations at the finer scale in the local temporal reality, where ideas about the space (i.e., mental patterns) translate into space. Later, due to interactions, competing and information exchange, only some of these mental patterns succeed in overcoming the initial mental patterns and their respective configurations. The diagrams in Figs. 6.5, 6.6 and 6.7 represent the winning archetypes that emerged from the competition between individual behaviors promoted by the main motive and the range of preferred geometric configurations and dimensions according to the level of hierarchy. As such, these diagrams represent the materialization of the "concern for privacy" and the avoidance of "visual corridors" (Kostov 2003, p. 63) by refracting, bifurcating, and closing perspectives with the aim of concealing life in a visceral space. Once translated into a shared experience, archetypes such as the *cul-de-sac*, narrow-angle solutions, and wide-angle solutions were transmitted as a modality or social norm from the top down, and space started to weave itself accordingly. Figure 6.8 illustrates how *cul-de-sacs* and "secrete rooms" act as a series of filters and feedback mechanisms from the top for accepted shared behavior. This process is further described in Sect. 6.4 on the complexity approach in terms of signaling, processing, and confirm–disconfirm of archetypes/configurations through levels. Further, as seen in the next section, this modality applies through scales, from a cluster of houses through to urban vistas. Once established as a social norm, there is a gradual translation into anthropological principles, intrinsic values, and the DNA pervading the unbroken web of wholeness, preparing the modality to reappear and germinate in another space–time within the web.

Figures 6.5, 6.6 and 6.7 show the repertoires of wining archetypes; the process of transition from mental space to the creation of physical space, as well as the geometric configurations and ranges of the dimensional preferences emerging from the competition of several individual mental patterns, or associative memories, and their respective configurations, promoted by the main motive. The process starts from the bottom (see also Figs. 5.2 and 5.3).

Figure 6.8 shows in detail how space was enclosed in relation to the motive of privacy, where a series of transitory spaces was interconnected, but at the same time followed strict rules in relation to the level of privacy and interaction. *Cul-de-sacs* and semi-private streets emerged as the winning geometric configuration or archetype.

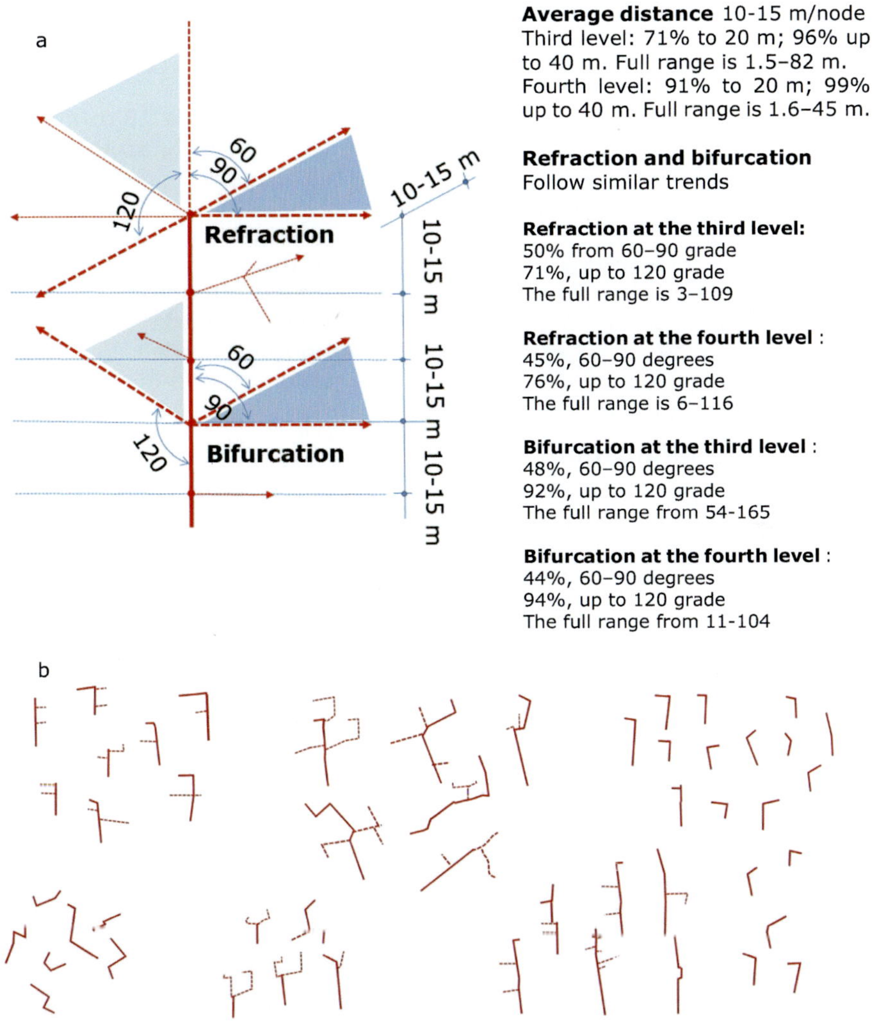

Fig. 6.5 Repertoire of wining archetypes at the bottom of the system. **a** The top figure shows the preferred dimensions and measures for the wining configurations below (as measured in sample 1.1 and shown in Appendix A). **b** The bottom figure shows the typical wining configurations at the bottom levels (family and group) resulting from the competing of associative memories in a bottom-up self-organized process, and feedback of wining archetypes at the upper levels

According to Portugali (2000, pp. 57–58), this would be a mental pattern recognition process. Priority in this filtered access was given to more consolidated structures, and their use depended on some of the oldest and well-off families that owned or controlled their adjacent access. Often, this was connected to private gardens to which access could be allowed or not (e.g., only some people or families could pass through these spaces). These material and immaterial "doors" acted as a filter and feedback

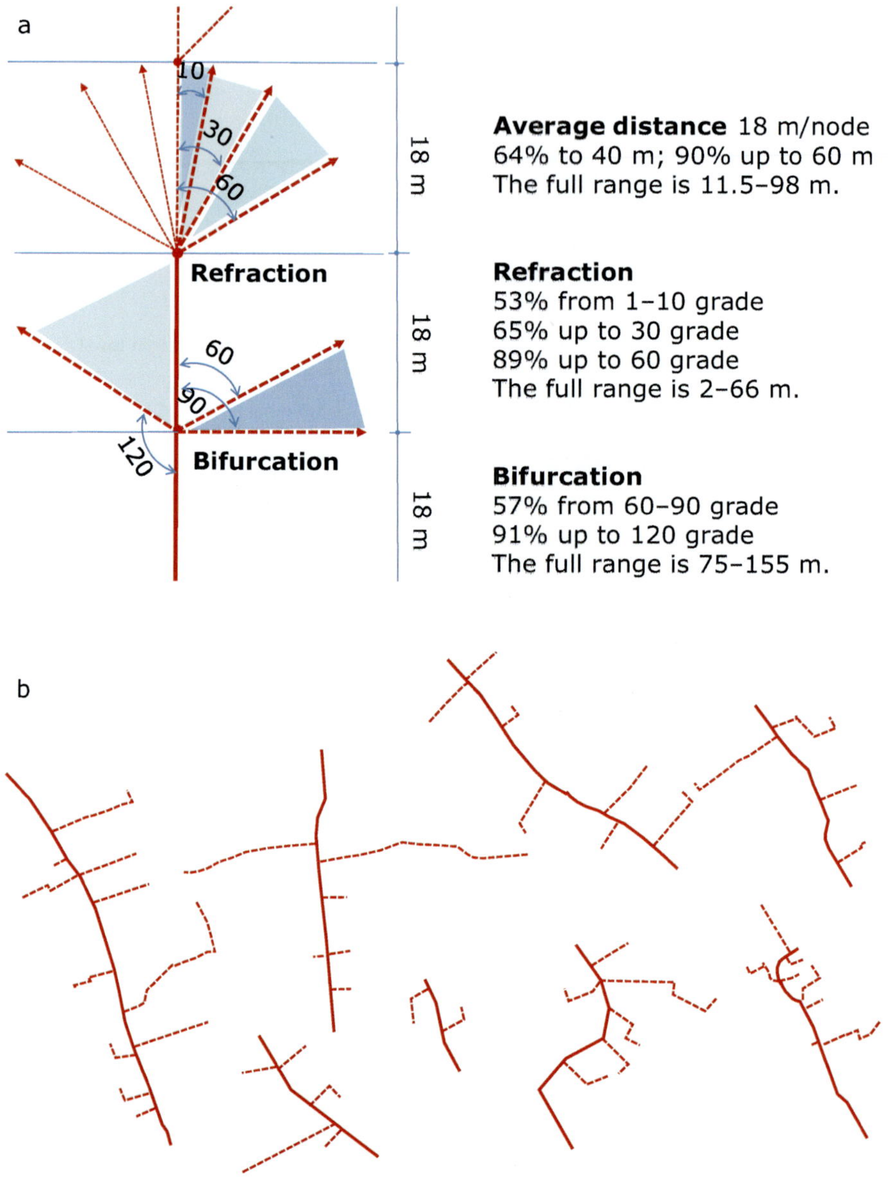

Fig. 6.6 Repertoire of wining archetypes at the intermediary level of hierarchy. **a** The top figure shows the preferred dimensions and measures for the wining configurations below (as measured in sample 1.1 and shown in Appendix A). **b** The bottom figure shows the typical wining configurations at the intermediary level resulting from the competing of groups (units of elemental space) to set the predominating shared behavior and configurations, and feedback the wining archetypes at the upper and lower levels

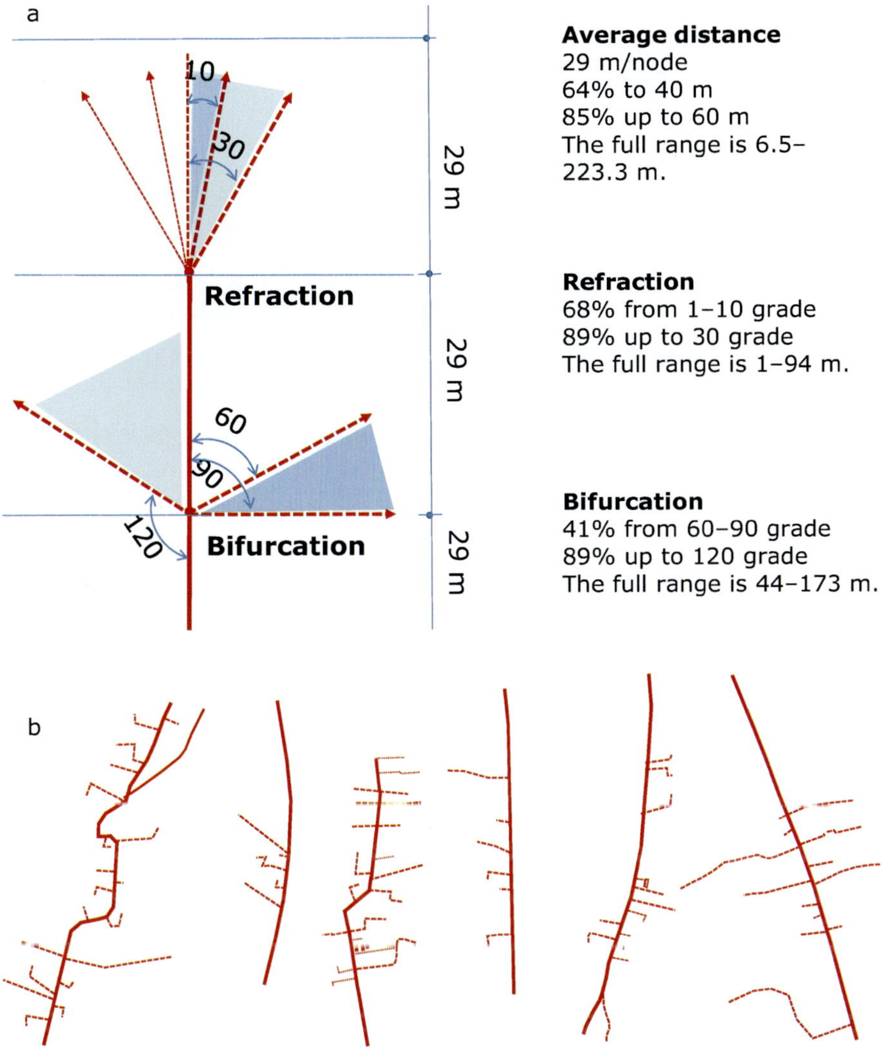

Fig. 6.7 Repertoire of wining archetypes at the upper level of hierarchy. **a** The figure at the top outlines the preferred dimensions and measures for the wining configurations below (as measured in sample 1.1 and shown in Appendix A). **b** The bottom figure shows the typical wining configurations at the neighborhood (upper) level resulting from the competing of groups (subsystems) to confirm the stronger rules and respective configurations that are feedback at the lower levels to order the entire system

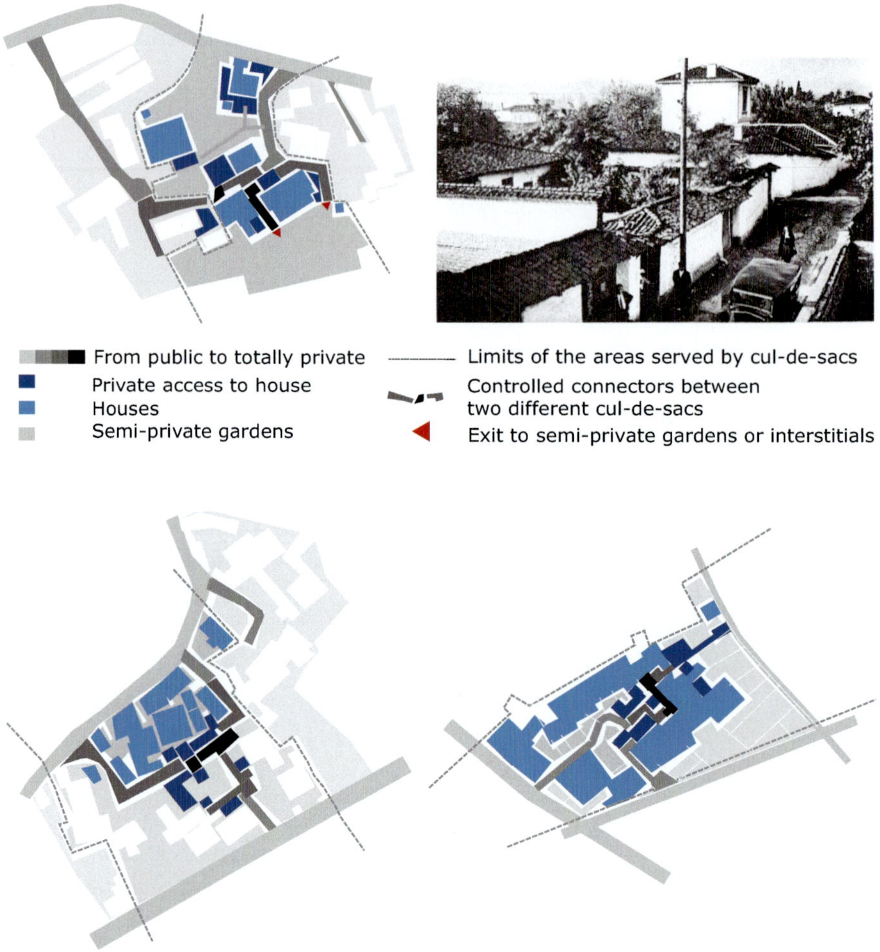

Fig. 6.8 Filters and feedback mechanisms for accepted behavior. *Cul-de-sacs* as a series of filtered or "secret rooms" to Eden and as an expression of the anthropological principles and visceral quality of space. The materialization of the right to privacy through physical filters and perceptual borders or perceived limited access (i.e., immaterial aspects). Public–private as a nuanced continuity

mechanism for accepted behavior. Access thus depended on following accepted rules and membership of certain groups. To balance this limitation, most houses had two or three means of access, or at least the potential for such. Easements in private gardens were a method of ensuring access and continued internal communication. As seen further in this chapter, filters play an important role in defining a modality as accepted social behavior. They also influence the interaction between levels of hierarchy in the system, allowing only some signals to pass between the upper and lower levels. Based on these similar principles, the entire city was reproduced (Fig. 6.21).

Returning to the events and event horizons shown in Fig. 6.3, while the insertion of high-rise typologies was a formal top-down process rather than self-organized, the winning archetypes are nonetheless those of the Historical Organic pattern, which controls the behavioral logic of the new typologies. This is visible in the planned layout of the constructions and their correlation with their surroundings.

In fact, this is happening within the web of wholeness, where people, through their imprint, are playing as agents maintaining the filaments of the unbroken web. This web depends not only on the permanence of the physical elements verifiable in the existing fabric but also on cultural and anthropological elements (e.g., conscious or unconscious obedience to rules and community problem-solving by entering into informal governance arrangements). This web was stronger than the urban design regulations that tried to impose regular patterns of land consolidation. This further illustrates that, before being a form in space, the labyrinthine structure of the organic pattern is a mental construct, a strong imprint. This is how people's imprint on space becomes transferable. They behave like an interfering pattern transcending time and space: interacting simultaneously with different places and times through their memory stores. This is how people adapt and re-adapt their environment in the natural city; that is, by bringing in non-material aspects or different cultural dimensions materialized in arrangements that gradually transform the environment.

Other Immaterial Aspects Within the Historical Organic Pattern

Perceived consent as energy: As mentioned, organic cities are self-regulated based on uncontested principles informally recognized by tradition that create the perception of what is allowed. This consent is like a potential horizon of the energy field, generated by continual negotiations of what is perceived as allowed within the system. However, this horizon has limits based on tradition. The energy of negotiations creates a propensity to change and for potential shifts from one status to another within established and consolidated traditions and existing building structures. Indeed, without the influence of an undisputed tradition as the associated wave factor originating from a social structure based on family and social control to balance the energy of perceived consent, this kind of system would fall into disorder. This perceived consent shifts up and down the system in the form of information exchange at the individual and group levels (arrangements), guaranteeing various degrees of self-regulation and temporary equilibriums. Situated between the rights of individuals to do something and their actual actions, perceived consent acts like a shock absorber transmitted from the top (i.e., the social norm). Conversely, within the limits of consent, the unpredictability characterizing this system makes it similar to open systems, which are based on the capacity of self-organization.

Intruder–guest transition: The energy of the intruder–guest transition becomes manifest when an outsider is trapped in the public–private continuity or uncertainty, as seen in the Historical Organic areas (Fig. 5.3). The public–private transition is mostly marked by perceptual and psychological barriers, rather than physical ones. In this case, the potential horizons of consent that grant or withhold access generate a sense of anxiety as we shift from the status of the intruder to potential guest, between a feeling of guilt that impedes us from continuing further and the attraction of the unknown. The pressure of uncertainty is reinforced by confusing signals of unexpected transitions from inner to outer qualities of space. For example, on a public street, one might see clothes drying (Fig. 2.1) or slippers left on the side of a door, while shops, bars, and other public services are often housed in private gardens or internal housing spaces. Decisions to penetrate further are made under a psychological sense of insecurity and with continuous interpretation of signals, such as sounds, smells, views, and atmospheres. It is based on these trial-and-error processes guided by sensations that the degree of admissibility of territorial shifts from one state to the other must be classified. Further, this admissibility varies across day and night and is influenced by the different layouts, such as housing composition and window–door orientation in relation to the public space (Arida 2002, pp. 188–189). These elements and feelings demonstrate that the Historical Organic is a highly adaptable and intelligent space that influences and controls the behavior of both residents and visitors.

This indeterminacy or double identity, which is directly related to a largely accepted formula for the use of space, is often perceived as a kind of freedom, albeit one that creates ambiguous sensations and an organically nuanced model, in which parts of public and private space are interchangeable. This flexibility, duality, and often "inefficiency" of the Historical Organic is an example of how the dichotomy between public and private can be complementary and nuanced, such that a space can be simultaneously public and private; that is, possessing a dual nature.

6.2.2 Recording Over as a New Emergent Reality

Original event: property as an entity in itself, or the frustration of being denied and/or having property seized by violence
Motive: re-appropriative revenge (associated wave effect of the event in the local temporal reality)
Behavior: inward (interfering pattern in the local temporal reality; at the bottom)
Modality: wrapping (as a group behavior, interfering pattern from the top)
Essential quality of space: involute (the emergent event horizon/reality)

In this section, I explore from a quantum perspective the holistic interrelationship that generates the Recording Over pattern, with its involute quality of space and related human properties. The analysis will consider the correlation factors of motive, behavior, and modality and discuss any similarities with the previously discussed Historical Organic pattern.

Recording Over has its historical and social roots in the original event that generated the motive for this pattern; that is, the denial of property rights after the Second

World War. This event combined with the weak public control after the collapse of the dictatorship, and the extremely poor economic conditions that accompanied the end of the dictatorship. In addition, perceived consent from authorities motivated a process of deregulation or, from a different perspective, triggered an organically self-organized process from the bottom. This process, described in Chaps. 2, 3 and 4, involved the continual subdivision of collective space based on multiple fragmented decisions coordinated at the smallest level, as an act of revenge against the collectivized space, imposed by the previous regime, that was supposed to incubate a new social life.

This revenge, which at its heart was a self-renewal of power by people to re-appropriate the emptiness surrounding them (Fig. 6.9), was triggered by the associated wave factor that emerged from the original event in which property was seen as an entity in itself, and people were frustrated about having been denied and/or having property seized by violence. As mentioned with regard to the Historical Organic pattern, the central unit of this society founded on tradition and ownership was the family. This idea lay in wait, only to reemerge as a powerful wave source of the motive to re-appropriate the urban environment, previously sterilized against social meaning, identity, and human properties.

This motive of re-appropriative revenge that pervaded the space triggered the individual inward behavior at the bottom of the system. The condition of concealing within the walls of frames and the potential to use the internal emptiness as a much-missed visceral space revived a new version of introvert behavior as a latent building block of the human brain. It arose as an interfering pattern, reactivated from a different space–time and social reality, that acted and overlapped with the preexisting building frames and free spaces. The result was a return to the organic principles of occupying the land with disregard for public space. This changing balance in the use of public space has been cyclical in Tirana. However, this new introvert behavior was born and forged within the framing walls and transformed into inward behavior. The inward behavior, triggered by the motive for property rights, was nourished by an unclear and uncertain environment. It progressed stealthily, first corroding the space from the inside and then from the outside through infilling and/or parasitic structures, depending on the relation with the frames, the functional and formal typologies, and the origin and degree of vindication (violation used in the past).

As the first signs of this new emergent reality appeared, each became a new source or potential event, influencing further developments through their own associated wave factors and interfering patterns. This behavior at the bottom self-regulated the infills and parasitic internal structures that gradually started to cyclically reproduce and repetitively control the same “wrapping” behavior at the macro level of the system, as the horizon of an unexplored potentiality of space made visible by the re-appropriative motive (associated wave). Thus, this inward behavior of re-appropriation, competing at the individual level, transformed into a shared experience or group behavior: the wrapping modality. This modality, in turn, dominates behavior at the macroscopic level, acting from the top of the system to control and guarantee the involute quality of space as an intrinsic value or as part of the DNA regulating the process within the system, which becomes transmissible within the unbroken web

Fig. 6.9 The emergence of re-appropriative revenge in the web of wholeness. Artwork by Endrit Marku, especially designed for this book

of wholeness. The process of wrapping the space and/or buildings from within and without consists of two main actions: wrapping buildings through attached and/or carved parasitic elements (which tend to osmose the original structure) and wrapping the free space within frames through the inward implosion of ubiquitous infills. The frame and free space re-engraving emerging from this process tend toward its maximal horizon (the territoriality of the event): the ultimate osmosis of the elements of Recording Over with the space and frames within which the process occurs, often creating a tense coexistence. Sometimes, the newly emerged re-appropriative paths (or recorded over shortcuts) flourish from the new uses, obstructing or even enslaving the preexisting ones.

The Recording Over pattern continues to act as an event, shaping developments (horizons) in the internal parts of Tirana. From this perspective, Recording Over is a new emergent reality or potential horizon of the previous land subdivisions or even of the event of the Historical Organic pattern itself (Fig. 6.10). Further, the transformed introvert behavior can be seen as a reactivated interfering pattern with a different identity; while the Historical Organic's refracting modality, which sought to avoid and conceal, transformed into an inexorable wrapping process within the frames (themselves considered intruders embodying different principles within the web of wholeness), with the aim of suffocating and seeking revenge on the collectivized spaces and recreating a new essential quality of space: from outwards to involute. The involute quality of space is similar but not the same as the visceral quality of space.

This newly emerged horizon already existed within the web of wholeness but became a tangible reality with its own identity and wave–particle aspect after the 1990s when the changed social conditions enabled the associated wave (e.g., property claims, revenge on unused public space, and weak government structures). In this sense, the Recording Over pattern can be considered a variation of the Historical Organic pattern. Importantly, the organic deregulation of the Recording Over pattern, as an emergence of the natural city, transforms the public–private dichotomy into quantum public private duality and allows for the flourishing of nuanced and complementary uses, as compared to a strict division of functions. These factors imply more complex physical and social governance arrangements.

The exploration above has sought to attribute meaning to the involute quality of space (i.e., the observable emerged entity in local temporal reality) as a formal expression or horizon arising from the associated wave effect (the motive) of the original event. The correlated conditions generated by these interdependent interests cannot be interpreted using linear logic. This is another step toward understanding the human properties of the natural city, with the goal of applying this understanding to the generative design process.

Repertoires of Motive—Generative Apparatus and the Genes Beyond the Patterns

The schematics presented in Fig. 6.10 show Recording Over as a transformed horizon of organic land ownership (the event), emerging as a new reality, and how this malleable field energy may evolve. They show the progression of the process, which

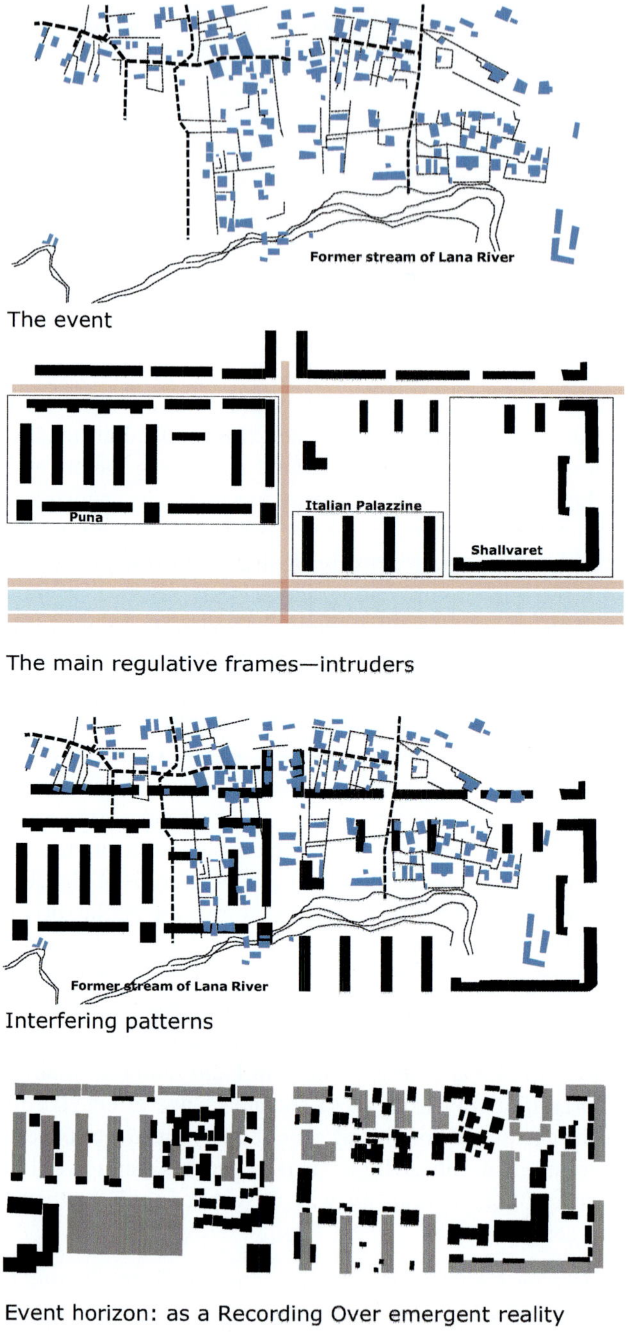

Fig. 6.10 Event–event horizon; recording over as a potential horizon of the previous land ownership, or of the historical organic pattern acting as an event

has tended to redefine the existing frames and spaces within them into involute enclosures; that is, into self-regulated sweeping and winding structures.

The diagrams in Fig. 6.11, viewed in combination with Figs. 5.25, 5.26 and 5.27, represent the repertoire of basic archetypes used to generate the involute quality of space. This is elaborated further in the diagrams in Fig. 6.12, which shows how the re-appropriation of public space materialized from a human motive, or associated wave, to create the involute quality of space in the local temporal physical reality. This occurs through a series of complex relationships in a self-organized settlement process (as explained earlier in this chapter, as well as in Sect. 6.4 below), as shown in each of the six steps in the figure in relation to pattern recognition, a process that occurs in the human mind.

6.2.3 *New Organic as a New Emergent Reality*

In this section, I explore the holistic interrelationships that generate the New Organic pattern and their related human properties through the categories of motive, behavior, and modality. For this pattern, the quality of space can be either collapsed in on itself (i.e., compressed) or inflated. The similarities in the generative processes between the New Organic and Historical Organic or Recording Over patterns will also be discussed.

Compressed New Organic

Original event: house-neighborhood, protective "fortress," *Rozafa* Fortress myth
Motive: isolation and amassing (associated wave effect of the event in the local temporal reality)
Behavior: densification (interfering pattern at the local temporal reality; at the bottom)
Modality: compression (as a group behavior, interfering pattern from the top)
Essential quality of space: collapsed in on itself (the new emergent horizon/reality); Fig. 6.13

As described in the previous chapter, the need for housing on the one hand and the risk of eviction on the other motivated the rise of massive and dense settlements, usually called informal settlements, where people grouped to face a common problem. The Compressed New Organic pattern was thus motivated by isolation and amassing, stemming from an associated wave factor that emerged from a combination of events centered on the need for dwellings as a natural human activity and the basis for life and frustrations over being denied dwellings, which emerged in the

Fig. 6.11 Basic archetypes (parasites and infills) that fulfill the requirements to generate the involute quality of space. See also Figs. 5.25, 5.26 and 5.27

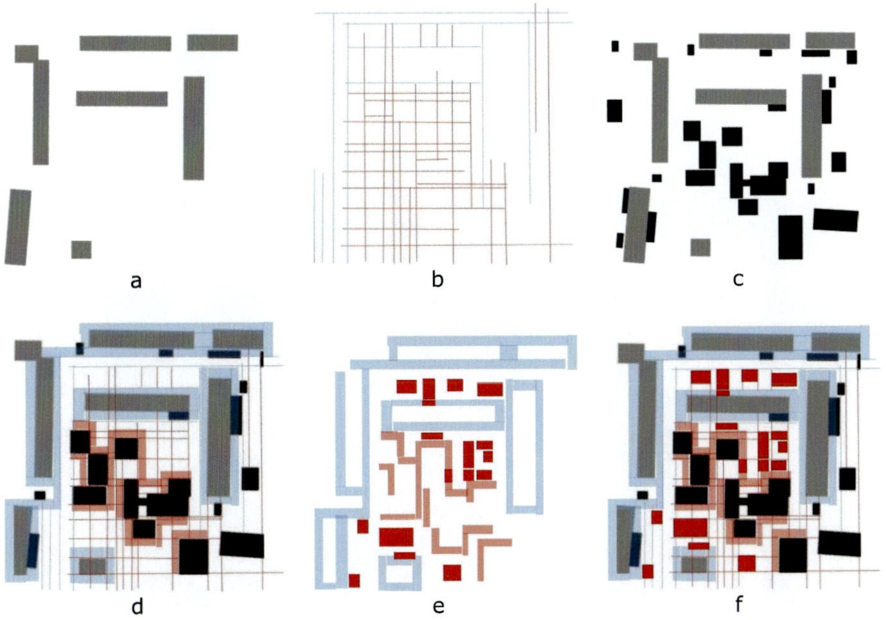

Fig. 6.12 The revenge on public space materialized from a human motive to the involute quality of space through repetitive behavior. Elements and geometric configurations at the finer scale in the local temporal reality, where the idea about the space translates into space. Specifically, the figures show: **a** centrally regulated frames and the seizure of private space; **b** individualization/atomization of the forcedly collectivized space; geometric configuration of a re-appropriative mental frame (pattern recognition): property as an entity in itself (the event) denied and seized by violence; potential space for acting; an associated wave of the event and interfering pattern; **c** materialization of this imprint (mental frame) in a temporary equilibrium of the recording over structure as a temporary horizon; **d** recording over of potential corridors transformed in a modality (collective behavior) of wrapping; if a new family occupies public space, they try to comply with these corridors (elements of governance based on group arrangements); **e** wrapping modality in potential further evolution; **f** essential quality of recording over as a new emergent reality; frames and sterilized space in full decomposition or re-composition; recording over in a more matured status of equilibrium: near to reaching its ultimate aim of the complete osmosis of recording over elements

specific historical and anthropological background of a series of interdictions during the period of dictatorship. However, given the circumstances, the event went beyond seeking shelter. Perceived consent from authorities paralleled people's fear of eviction by the state. From this fear emerged a collective stratagem to first isolate and then amass; that is, to auto-protect or "fortress size" before collapsing. This condition created the specific enzyme for space that could collapse in on itself to protect the built environment when absolutely necessary (Fig. 6.14). Thus, this event is not just a house; it is a protective fortress—a house-neighborhood.

At this point, it is necessary to clarify the complementary dual couple of collapse and protection (resistance). What kind of collapse are we speaking about? The space, represented by house, "need to collapse" to sublimate and overcome the issue of

Fig. 6.13 Isolation and amassing to recreate protective fortresses. Artwork by Endrit Marku, especially designed for this book

survival in a hostile environment. This was a new meaning of space. Thinking of the house-neighborhood and protective "fortress" as the main event of this pattern, it can be considered the horizon of another non-local temporal event within the web of wholeness and with a different nature, generated in the mythological dimension: isolating, amassing, and collapsing in on itself can be viewed as parallel to the myth of self-sacrificing to reach a final aim. The original event of this myth can be traced to the *Rozafa* Fortress, a commonly known legend in the Balkans, in which death is used as a measure of infinite time. In the legend, *Rozafa,* a young woman and mother, had to be walled up in a fortress so that it could stand. Thus in our context,

Fig. 6.14 Fortress sization to protect the built environment, 1994 (credits, Babani)

under the omnipresent danger of eviction, death or collapsing in on itself (because of people's amassing—expressing the myth of sacrifice) is used as a warning pressure and menace against authorities in order to prolong the life of man-made artifacts. Collapse as a warning and menace is an attempt to synergize the attention of the forces of the universe (here, administrative authority) to make things (dwellings or fortress) last, ideally forever.

Isolation as a recursive theme of the natural city, and sacrifice as an event that aims to influence the perception of time or extend its horizon indefinitely, are key components in the specificity of this pattern. In fact, in real cases of eviction and demolition, people were ready to sacrifice themselves together with their home, strongly paralleling the legend of the *Rozafa* Fortress (Fig. 6.15); however, in the end, these people were driven out by the police. This raises the question of whether the administrative authority and the law enforcement can work together rather than in conflict, negating any requirement for extreme sacrifices. This is the reason for a motive that, while seemingly strange and inhuman, is the materialization of the associated wave from a concrete and intelligent local temporal stratagem.

The condition created by the motive of isolation and amassing that pervaded the system triggered a pressurizing process that translated into densification as a behavior at the individual level or an interfering pattern in the local temporal reality at the bottom of the system. Under the conditions of rapid urbanization, this behavior could be seen as a tendency toward mass compactness, born and forged under the pressure to reach a critical mass as the first step toward auto-protection (Fig. 6.16). The concealing of this kind of individual behavior within the mass can be seen as

Fig. 6.15 New *Rozafas*. Evictions and the protection of the fortress as an event existing in the mythological dimension. Self-sacrificing and walling up to strengthen fortress-houses against the forces (law and government) that seek to destroy them

echoing introvert behavior, transformed under the pressure of uncertainty by the need to densify and remain together in a condition of forced nearness.

Fillers or densifiers are the basic archetypes used to generate the collapsed-in-on-itself quality of space, which constitutes a forced nearness or fortress sization (characterized by auto-protection, survival, and sacrifice) and a mental pattern recognition of the original event (i.e., the house-neighborhood-fortress). In this way, isolation and amassing materialized from a human motive, or associated wave, to the collapsed-in-on-itself quality of space, representing the space of surviving in the local temporal physical reality. The occasionally created dead ends are not, as was the case in the Historical Organic pattern, interconnected by secret transitory rooms acting as filters of control for social behavior; rather, they exist in a condition of forced nearness and express tension, fear, and inefficiency in a vague space. Separate access is needed for each house.

The fact that people in these neighborhoods often do not belong to the same area of provenience or have the same culture (including religion) makes it difficult to adopt a common social formula on the use of space. Instead, tradition as an amalgamating factor is replaced by perceived social injustice and uncertain survival, binding people together and becoming an important reference for all. Thus, the lack of a unified identity represents a significant point of contrast between how

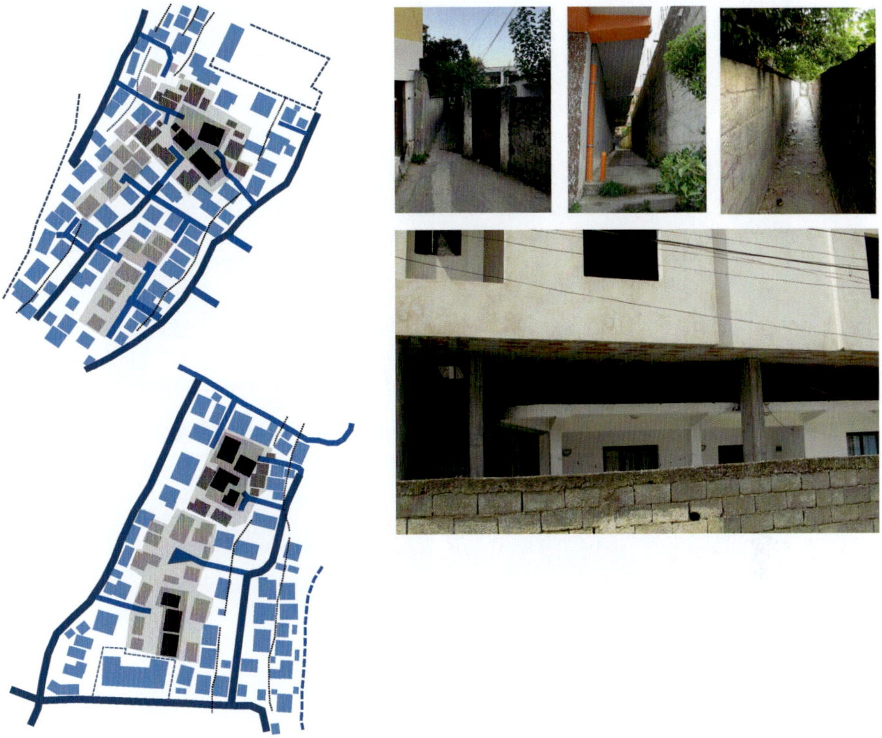

Fig. 6.16 Event horizon of the fortress, showing the walls raised under the pressure of compression and densification. Buildings are either fillers or densifiers (creating the mass), but in the clashing core (marked with gray), they become "colliders," sometimes even combined in collision rows

the Historical Organic and the Compressed New Organic areas were shaped. This cultural clash is also manifested in the incoherence of the architecture (Figs. 5.40 and 6.17); however, the apparent lack of coordination in these images is misleading. In actuality, individual arrangements and information exchanges were being intensively carried out at the level of pragmatic human-scale concerns, while different interfering patterns overlapped within relatively small neighborhoods to create potential tensions (i.e., different events and the overlapping of different event horizons). Often, external support was needed, at least initially, to improve coordination at the bottom and higher levels of the system (i.e., the neighborhood) under these conditions of forced nearness.

This kind of settlement behavior at the bottom of the system self-regulates the level of density increase, or the internal structure, which gradually starts to dominate and cyclically reproduces (as an order parameter) the same behavior at the macroscopic level of the system through the modality of compression as a shared behavior. Transformed into a social pattern, the modality of compression dominates the process from the top and moves infinitely close to the collapsed quality of space without ever

Fig. 6.17 The competition between different mental patterns (memory and information from countries of origin) and alternation of temporary equilibriums remains dynamic even two or three decades since the inception of this pattern. The resonation of different mental patterns gives rise to new emergent realities. Collapsing in on itself represents both the continuous risk of losing the precarious equilibrium and the potential to transit to a higher degree of self-organization by borrowing energy from the environment. In this scenario, there is a sensation that everything is unfinished and in a tentative state of becoming something different from what it is

collapsing (being controlled by this modality). From this pressurized organic aggregation, houses and other activities are in a special condition of nearness, close to a clashing condition, that simultaneously reflects the extreme and pragmatic efficacy in the use of a formula for a space based on surviving. This collapsing in on itself as a new emergent reality represents a quality of space generated under high pressure and an unfriendly environment.

To understand it better, we can compare this quality of space with that of the two previous patterns. The new horizon emerging from the collapsed quality of space differed from the visceral quality of the Historical Organic and the involute quality of Recording Over and was strongly influenced by the high pressure of eviction.

From this perspective, the collapsed-in-on-itself quality of space may be considered a pallid new emergent reality or different horizon of the event of the Historical Organic pattern. Similarly, the introvert behavior of the Historical Organic is transformed in the Compressed New Organic into densification behavior, which here implies an anonymizing process (concealing behind) forged under pressure and by uncertainty. This is a reactivation of concealing as an interfering pattern, but with a different identity. The refracting modality of avoidance in the Historical Organic became the compression modality. In this modality, the possibility to refract in extreme situations may not exist because the space is compressed to the point that it could collapse in on itself. Thus, no refraction is possible without breaking the space. Under this new modality, the right to visual privacy as a mental pattern reaches its maximum blindness: space becomes meaningless, distorted, and deflated, devoid of clear anthropological factors. The space comes close to ceasing to exist (sacrificing itself). However, as with the generation of stars, it is in this collapse that the peculiarity and potential to generate new emergent realities from the same pattern exist. The motive contained the idea of collapse since its initial conception.

As mentioned in the previous chapter, despite the unfriendly environment, there are also instances of a more natural co-evolution of this pattern, resulting from agreements between different forms of governance, including local residents, formal institutions, and international organization. Figure 6.18 shows some of these areas two to three decades later since the initiation of this pattern. This kind of co-evolution based on "nuanced and contextual" (Suhartini and Jones 2019, pp. 14, 220) urban governance illustrates the need to keep the system of interrelationships open. For this, planning and design need to reflect from a holistic perspective the system of interrelationships underpinning the real life.

A final question remains: what are the human properties of this pattern? At first glance, there appears nothing to extract from this generative process centered on the condition of collapse. The process starts by creating the condition of neutral nearness, which soon evolves into forced nearness or even forced intimacy, potentially producing social conflicts that risk causing a condition of collapse. As such, this pattern can be considered an *in extremis* condition of the Historical Organic pattern. However, it also contains two important properties. First, it has the potential to reach an equilibrium of a near collapse condition without ever collapsing (the modality of compression). This property is an important device for generating equilibriums in a design process and is in evidence in many areas of Tirana, where an admirable state of equilibrium and consolidation has been reached. Second, the Compressed New Organic pattern contains the potential for a specific self-organized process that may drive self-destruction with the potential to reemerge. This may be an important human property in the design generation process, needed as an intelligent auto-destructive and regenerative device (see the further discussion about *autopoiesis* in Sect. 7.3).

Fig. 6.18 Captured two to three decades after the start of this pattern, these images show the compression process at the neighborhood level; the event horizon of the fortress; the near-to-collapse condition, reaching toward a stable equilibrium; and the secret of being near to collapse without collapsing. This condition emerged from interactions in a bottom-up self-organized process and co-evolved because of nuanced governance arrangements in an open system of interrelationships

Inflated New Organic

Original event: isolated house as the focal point of territorial units; widespread or scattered as part of extensive villages or communities
Motive: isolation in loneliness for protective reasons (associated wave effect of the event in the local temporal reality)
Behavior: dispersive (interfering pattern in the local temporal reality; at the bottom)
Modality: floating (as a group behavior, interfering pattern from the top)

Essential quality of space: inflated (the emergent event horizon/reality); Fig. 6.19

The Inflated New Organic pattern as a settlement model was motivated by isolation in loneliness. This is the associated wave factor that emerged from the correlation of original events, which differed slightly from those of the previous case, leading to a different quality of space. Many of the factors mentioned as triggering the Compressed New Organic pattern (e.g., the self-provided housing system, the activity of dwelling as an entity, the lack of affordable housing alternatives, and weak public control) are also the factors underpinning the generation of this pattern.

Due to the more homogeneous origin of the settlers in this case, who mostly came from the north of the country and who were consequently driven by more unified anthropological principles and exhibited a predominating culture, this settlement model demonstrates logical (although not formal) similarities with the Historical Organic pattern. Like the *kulliyes* of the Historical Organic pattern, the vernacular typology of the scattered and isolated dwellings described in the previous chapter (Thomo et al. 2004, pp. 27, 75, 97, 190), as the focal points of territorial units, is the event or predominant mental pattern in this case (Fig. 5.45 showing the original settlement model). This event generates the arousal effect of a settlement model motivated to create units of isolation through loneliness in space. This motive of isolation in loneliness triggered an organic, slow but steady self-organized settlement process from the bottom. At the individual level, pressure for coordination was minimal, as coordination and/or subordination was done with (and within) the elements of the territory.

While in the Compressed New Organic pattern, the strategy to auto-protect generated a common social cause, in the Inflated New Organic pattern, the strategy to auto-protect generated the opposite motive: dispersion as auto-protection to avoid public attention, to hide non-conformity with the law. This is because the associated wave was sourced from a different tradition rooted in loneliness and isolation in space. Thus, similar conditions in the local temporal reality produced these different qualities of space because of the correlation effects derived from the different event backgrounds. We can imagine these patterns as different branches within the same web of wholeness, germinating together but diverging after a certain point. This specific correlation effect within the web of wholeness produced the genius of a collective stratagem, or the main motive, to generate a quality of space that in extreme situations could inflate. With excessive inflation, the normal meaning of space is lost or even stops existing. Elements (e.g., houses and economic activities) within/constituting the space can “float” or “disappear” in superfluity.

The condition created by the motive of isolation in loneliness triggered a dispersive process that materialized as individual behavior in the local temporal reality. This behavior of individuals trying to avoid vicinity acted as an interfering pattern at the bottom of the system. The result was the conception of a model in which the house was an integral part of the territory and of all adjacent natural elements (e.g.,

Fig. 6.19 Isolation in loneliness as the focal point of territorial units. Artwork by Endrit Marku, especially designed for this book

rivers and forests), land and other resources of life were perceived as abundant, and there was an awareness of breaking the law and a fear of the controlling power. This locational preference to scatter or disperse within the intimacy of the natural or agricultural elements, marked by land subdivision and service routes, gradually became a shared behavior.

Even in the Inflated New Organic pattern, the system needs to reach a critical mass to achieve auto-protection. The pressure to scatter and diffuse forges a group dispersion toward a near-to-inflated condition of space, tending toward total evaporation or disaggregation of the material objects defining the space. This kind of evaporative behavior echoes the introvert behavior of the Historical Organic pattern, transformed under conditions of floating and uncertainty and the need to evade and disappear. This is an example of how, under the quantum approach, concealing, densification, and dispersion become complementary concepts manifesting similarity. In this way, even dispersive behavior contains echoes of introversion, with isolation reinforcing the survival of individual anthropological aspects. Moreover, even in the pattern, the condition of nearness may eventually appear, buffered somewhat by the long maturation time in this case. That the interaction process in the initial stages is based on different principles to those seen for previous patterns is the main value of this pattern.

This settlement behavior at the bottom of the system self-regulates the level of dispersion in the internal structure. This model gradually dominates and cyclically reproduces (as an order parameter) the same behavior at the macro level of the system through the modality of floating and fusion within the territorial superfluity. Here, fusion means that the house, as a focal point within the landscape, creates a unit system with similar characteristics through scales, from the garden to the open territory. Once transformed into a social pattern (common behavior), the modality of floating controls the process and moves infinitely close to the inflated quality of space without ever liberating the space from that quality. Thus, even when the number of houses increases, the same principle applies but with an infinitesimal logic. In this kind of dispersive disaggregation, houses and other activities are in a special condition of nearness, produced by the DNA of isolation deriving from the original event (the associated wave effect of the vernacular houses). This is a distancing based on infinitely reducible self-similar unit systems, as discussed in Sect. 6.3 with regard to the fractal approach. Most importantly, the units or modules reflect the same anthropological model: the relation of the house with the territory, which is specific in each case. This model of the unit system is mostly set by territorial specificity and the subordination of the houses in relation to their elements, rather than by coordination or individual negotiation, as in the other patterns (Fig. 6.20). This pattern reflects the advantage of the more unified anthropological principles and the efficacy of a logically infinitesimal distancing as an imprint in the use of space.

The Inflated New Organic, as a new emergent reality, represents a quality of space generated under a dispersive behavior, in an environment ignored by the central authority, where groups of houses, initially scattered in the landscape, become fused within the territorial superfluity. Specifically, the house, as the smallest element in the system, is part of an integral self-similar territorial system comprising units that

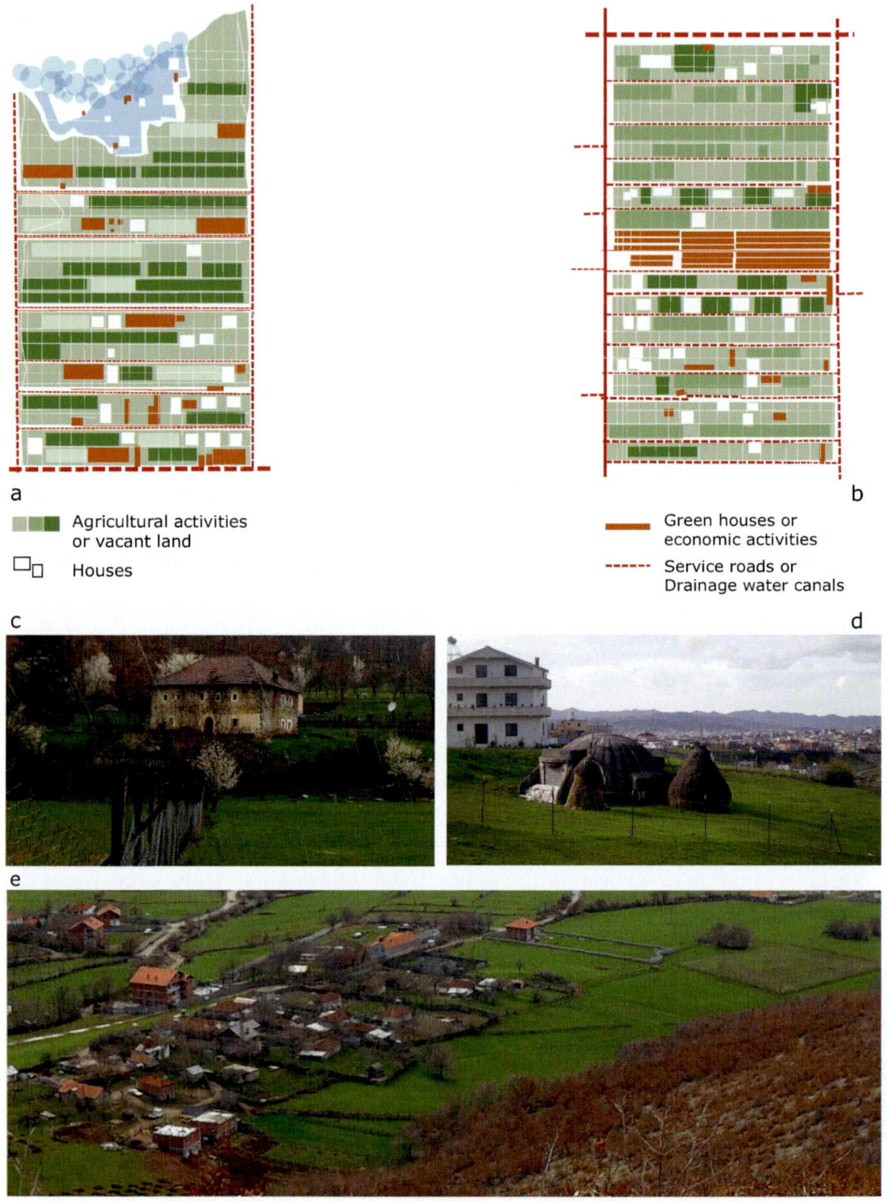

Fig. 6.20 The images show how the motive of isolation in loneliness, as the first condition to inflate the space, takes shape. This motive is a mental imprint of the original event (i.e., the house as part of a territorial system) and generates the mental building blocks by which the idea about the space translates into space (here, the house-territory unit system as the most used archetype to inflate the space). The top images **a**, **b** show the rationally inflated pattern, which maintains an infinitesimal distance that reflects the geometric relationships of the larger scale. The network is the guarantor of superfluity, or special nearness. The images in the middle **c**, **d** are of the house-territory, while the bottom image **e** represents how isolation as a mental construct maintains its quality independent of distance and the space available

follow the same anthropological principles through scales. As such, even a house's garden is a micro-universe or subordinated unit of the territorial integrity grouped around the house. Even in a denser condition and at a smaller scale, this quality of space is preserved. It transforms in a rationally inflated space through individual acts of maintaining infinitesimal distance, reflecting the same geometry or principles of the larger scale.

This new emergent reality may also be considered a vaporized version (horizon) of the visceral quality of space of the Historical Organic pattern, considering the latter as an event. It is a reactivation of concealing as an interfering pattern, but with a different identity. The refracting modality of avoidance of the Historical Organic transformed into the floating modality of elements in the system (i.e., infinitesimal detachment) where, in extreme situations, the possibility to refract would not exist, as the lines would be infinite while the space inflates. The right to visual privacy, as a mental pattern, if put under these new modalities, reaches its maximum extension. Further, space loses its traditional significance as an atomized entity with limits and borders because of inflation or the logic of infinitesimal superfluity. This close to dissolving condition generates the property of "maximal extension" (i.e., maximum things can fit in it). Under these conditions, the anthropological factor may be reinforced as a separate or isolated event in space (traditionalizing) or gradually diluted (de-traditionalizing) because there is no need for agreement. Here lies the main value and potential of this pattern to generate new emergent realities from these specific enzymes: infinitesimally inflated space.

This pattern initially appears wasteful, with nothing to extract from its human properties. The process starts with isolation in loneliness as the first condition to inflate the space. Interestingly, this condition is preserved even in the face of further developments, because the quality of inflation is a mental pattern before a physical expression. This special condition of nearness or distancing, created not by aggregation but by dispersive disaggregation, is the most important human property for incorporating into the DNA for creative reproductions. This pattern may be considered an *in extremis* condition of the Historical Organic pattern and an opposite extreme to the Collapsed New Organic pattern. Here, space loses its traditional significance because of the conceptual superfluity: street, square, and garden lose their meaning because of inflation. It is in these generous interpretative potentialities that this pattern offers its best. It contains the potential of a specific self-organization that may approach a proliferative self-inflation of space through scales, independent from density. This motive is self-contained and is an intelligent auto-proliferative device.

6.3 Fractal Approach—Patterns as Relational Order

For a more holistic understanding of the urban patterns identified in this book, this section presents the patterns and their representative samples as fractal systems with elements that irregularly repeat in a similar way through scales. The contention is that wholeness is a relationship that emerges through recursive behavior that creates the invisible structure of the form.

6.3.1 Historical Organic

In this section, I show that the site-specific irregularity of the natural city manifests fractal properties. This raises the question of how the essential quality of patterns identified above is distributed through a hierarchical structure, and whether sets of elements and subassemblies within this structure repeat according to similar principles through scales (hierarchy). From this recursive behavior emerges the invisible structure of form and its holistic dimension. Starting with the example of the Historical Organic pattern in Tirana, here I illustrate how the categories of the fractal city help to define the recursive irregularities characterizing the visceral quality of space (i.e., path deviation and nuclei isolation) and how they repeat geometrically through scales. This helps to elucidate the process cascade important to later discussion in Chap. 7.

Once we understand its organizing principles, we can study the behavior of the elements within the Historical Organic system structure, or the internal dynamics that create that form. As mentioned in the previous section, the patterning process is set into motion by the motive of the right to visual privacy, which triggers repetitive actions of introvert behaviors. This introvert behavior, as an interfering pattern, acts as the dynamic at the bottom of the system structure. This settlement behavior is reflected in the modality of refraction, as recursive or sequenced action, applied to the straight lines of the street (motion), to scale down or replace them. Thus, modality, as a social behavior emerging from competing principles, self-regulates the system from the top and creates the scalability principle of the pattern by ordering settlement procedures and the elements of space related to this (e.g., recursive deviation and plot subdivision). In this way, self-similar or self-affinitive physical forms are created through levels, pervading the entire system structure.

These internal systems of interrelationships constitute the dynamics that define the external form of the Historical Organic, materialized in its morphological expressions. This is observable in the recursive logic of the labyrinth (Figs. 6.21 and 6.22) and the isolation during the settlement process, which scales up and down in a self-affinitive way across levels of urban hierarchy.

As Batty and Longley (1994, p. 60) expressed, we can understand the order within fractals according to three main principles: they are self-similar in the degree of irregularity manifested in the whole and its parts; they can be described in terms of a "hierarchy of self-similar components"; and their "irregularity" in form creates "continuous" forms that are "nowhere smooth." Analyzing the urban patterns according to this logic, we obtain information not only about the structure of the system, but also about its behavior and the reasons behind its dynamics.

In the analyzed samples, the self-organized settlement process generated a self-similar structure and dynamic (behavior) of the system. Thus, from the single

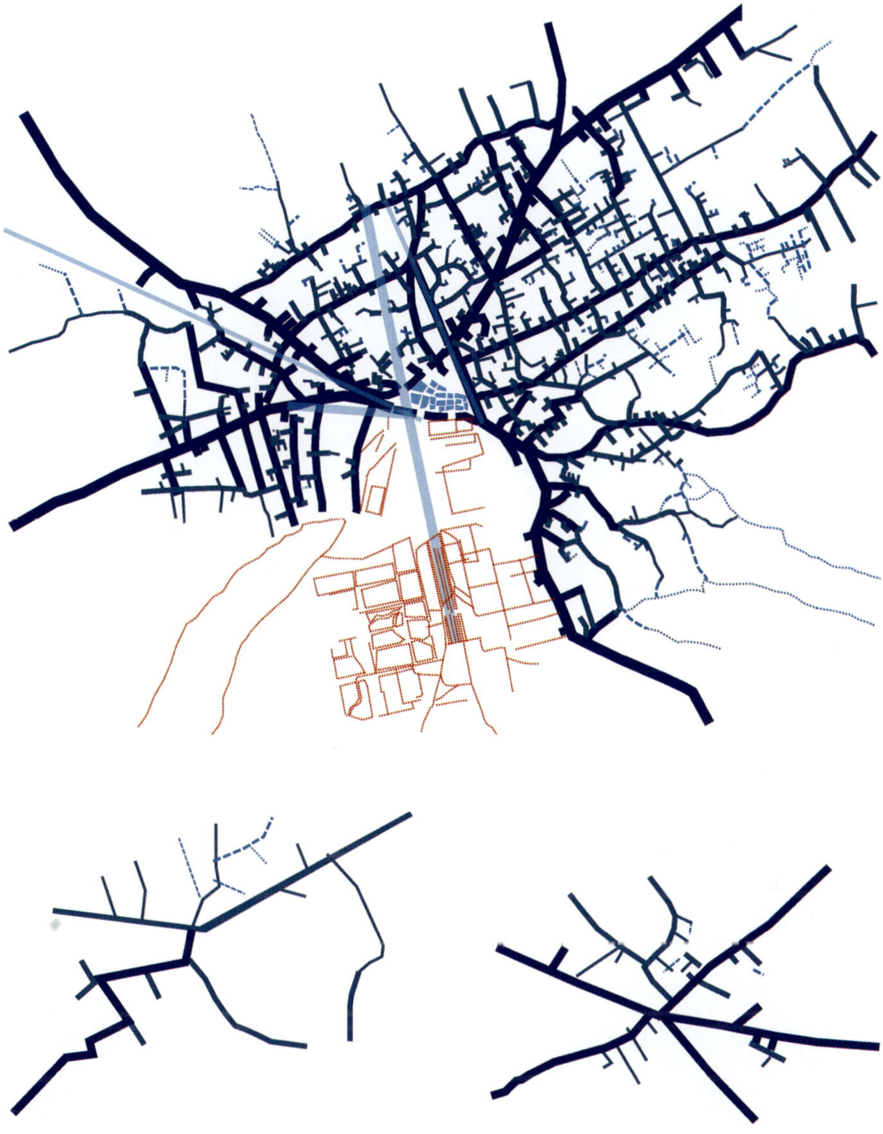

Fig. 6.21 Self-affinity as an organizing principle and irregularity that repeats geometrically across scales: from the city as a crossroad to an ordinary crossroad within the city. The city is a pattern inside a pattern reproduced on similar principles; see also Fig. 6.8 showing a series of filtered access points influencing interaction between levels of hierarchy by "defining" a modality as accepted social behavior

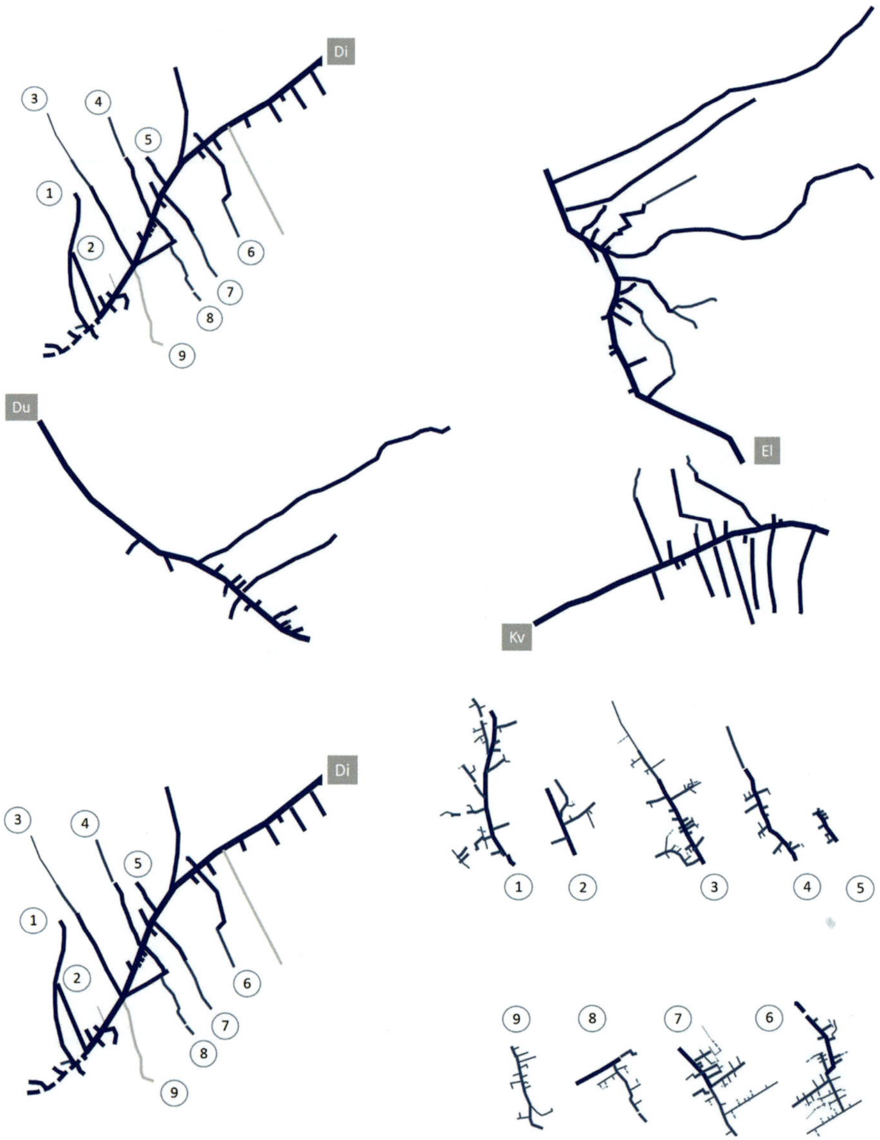

Fig. 6.22 Tirana's four organic generative paths. The refraction and bifurcation as recursive irregularity across scales indicate static (structure) and dynamic (behavior)

house to the neighborhood, there are at least four levels of hierarchy (Fig. 6.23) within the system (considering the neighborhood or sample as a system), ordered by the modality of path deviation, which is the dynamic controlling the macroscopic behavior at the system level. The elements at the bottom of the system, the lowest level at which this modality acts, consist of individual housing plots and

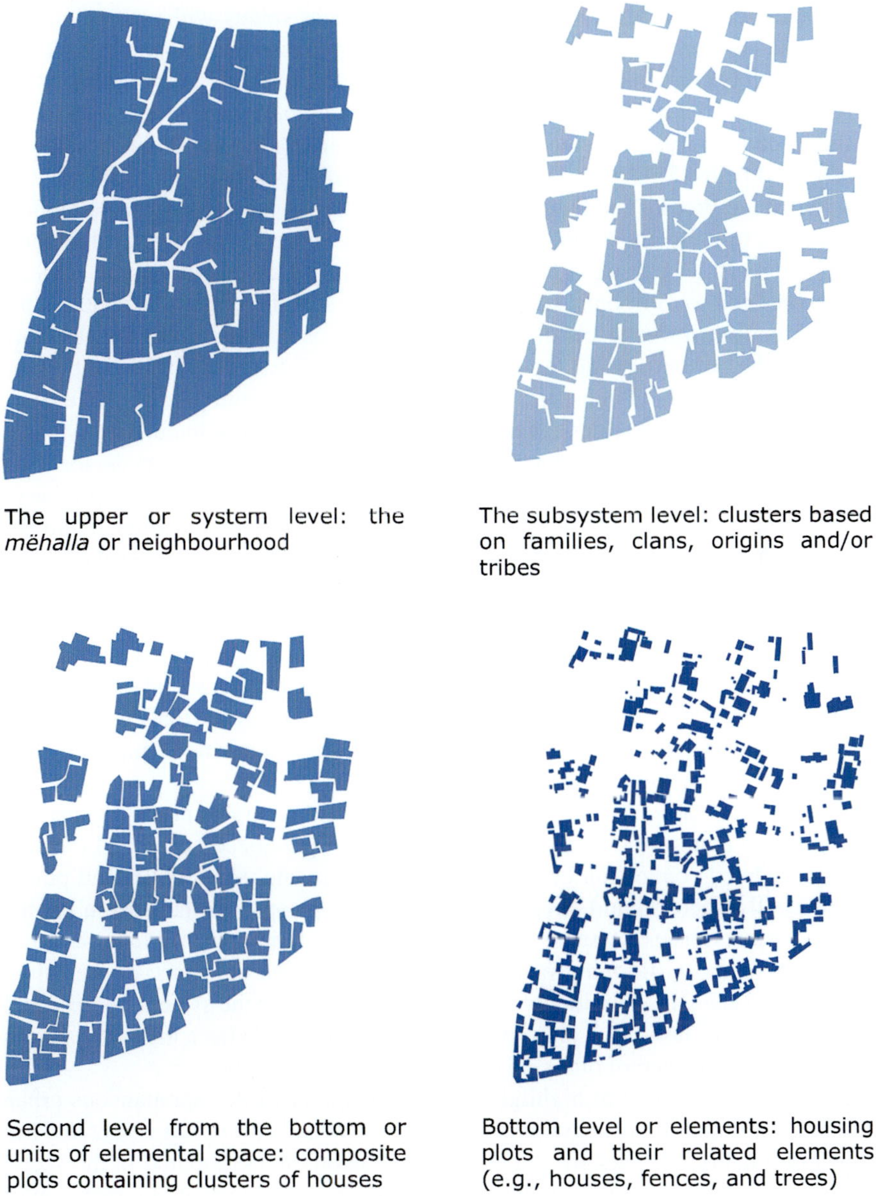

Fig. 6.23 Levels of hierarchy within the system: elements and subassemblies. This self-affinitive system structure applies the right to visual privacy in the settlement process

other related elements such as the house itself, access streets within the plot, and enclosures. The next level of hierarchy, or the units of elemental space, consists of composite plots containing clusters of houses based on family and other subdivisions, fed by the deviated access paths. The next level up, the subsystem, includes a number of clusters formed on the basis of families, clans, origins, and/or tribes, fed by the refracted paths. This process of group creation was especially important during the initial stages because it determined the modalities of the subsequent settlement process. The initially ample interstitials could leave space for further developments following the same settlement logic. The upper level of the hierarchy, the *mëhalla* or neighborhood, corresponds to the entire structure of the system. It contains the three consecutive lower levels described above, which correspond to the structure combining the bottom-up and top-down self-organized dynamics of the settlement process. If we extend our consideration to the city and its geographic container (i.e., the urban and territorial scales), we see a fifth and sixth level of the hierarchy.

Each of the abovementioned levels is characterized by, and guarantees, the right to visual privacy as the motive pervading the system, resulting in a self-similar or self-affinitive visceral quality of space through a hierarchical structure that manifests the whole in parts. From the house layout to the higher level, these levels respect the same principles. Thus, self-similarity is manifested as a continuous repeatable irregularity through a hierarchical structure. This logic does not depend on how much space is available or on any physical determinants (e.g., topography); it is the direct result of self-organization and improvisations based on a specific social structure and culture. Specifically, it is based on the anthropological persistence that emerges (i.e., the event–event horizon) at various levels of the urban and territorial hierarchy and is materialized in the urban morphology (Fig. 6.24).

What has been described above represents a typical recursive process generated by a principle that scales up and down to seclude a privatized space and create a socially accepted connection between private, less private, semipublic, and public spaces: ideally, there should never be a straight-line perspective between a private and less private space. As such, the deviating principle is not only manifested in the settlement and positioning of the initial groups within the space of the future neighborhood (i.e., the individual behavior at the bottom of the system, represented by clan or family clusters), it also appears at the upper levels (as a modality) and in the *mëhalla*'s avoidance of other neighborhoods, at least initially.

Finally, it is important to highlight that the geometry of this spontaneous urban order is not abstract. It occurs within a specific social and cultural construct and has its own rationality. It arises from a continuous negotiation embodied in the social structure and its informal agreements. To give meaning to the spatial dimension of subassemblies, from single elements through to units of elemental space and the system as a whole, we explored the anthropological principles as the building blocks (mental frames) that generate new emerging realities. These are part of the correlation effect within the unbroken web of wholeness and occur where quantum and fractal interpretations meet. This is a geometry put in motion by a specific meaning emerging from the web of wholeness. In this sense, the fractal approach addresses issues related to generative processes, including morphological expressions of human patterns.

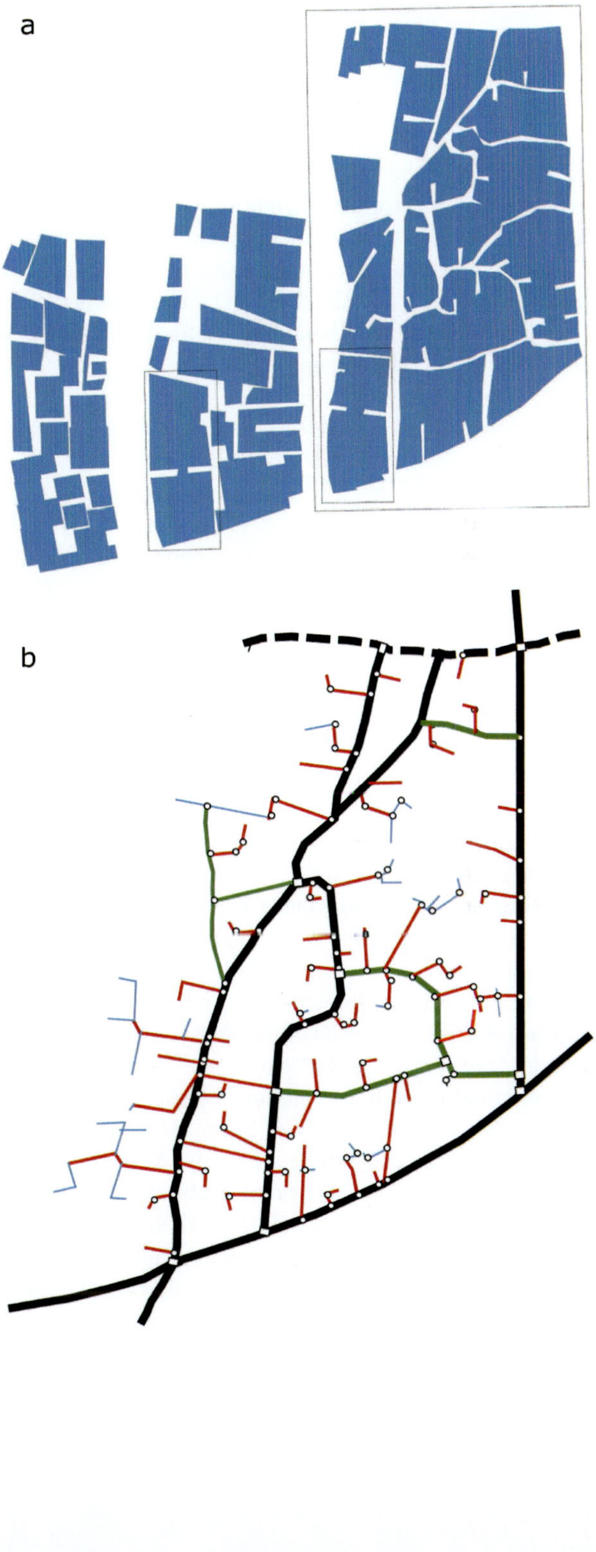

Fig. 6.24 **a** The whole in parts. From the system to its elements, a similar irregularity is manifested throughout, albeit on different scales. This reflects a continuation across scales, reflected in the subdivision of land. **b** The modality of recursive deviations across scales motivated by the right to visual privacy as a self-affinitive principle that scales up and down to generate the visceral quality of space

Motive, behavior, and modality are important concepts for understanding the underlying triggering factors that create the dynamics for the emergence of the recursive irregularities represented in the self-similar or self-affinitive system structure. They correlate the space with the human properties that create this space, reconnect material and non-material aspects contributing to the pattern generation process, and link the mind to the physical. As such, they are important triggering factors underlying the external form of neighborhoods and cities. Thus, redefining the visceral quality of space from the fractal perspective, we can characterize it as the process of creating and experiencing the space from within. This creates a kind of privatized space in which organic Eden(s) can be isolated as a continuous logic through scales, connected by refracted visual lines in a self-similar or self-affinitive logic, from the individual to the urban scale.

6.3.2 Recording Over

Continuing with the Recording Over pattern in Tirana, here I illustrate how a fractal understanding helps to define the recursive irregularities characterizing the involute quality of space and their geometric repetition through scales (i.e., wrapping through attached and/or carved parasitic structures or implosive infills). In this recursive behavior, the underlying structure of the external form of the Recording Over pattern becomes apparent. Considering the analyzed patterns and their morphological expressions as systems with structure, we first identify the sets of sub-assemblies constituting the structure and the recursive irregularities in this structure, before analyzing the dynamics of the system and the reasons behind it.

The levels of hierarchy in the Recording Over pattern are intimately related to those of the frames in which this pattern binds through parasitic structures and infilling. Therefore, understanding these hierarchical levels provides insight into both the frames and the elements of Recording Over, which are intimately related to those frames (Fig. 6.25). This helps to elucidate the dynamics of the system form.

From the smaller elements (e.g., parasites and infills) to the system as a whole, there are at least four levels of hierarchy. At the bottom of the system, as the smallest particles, are the elements; that is, the individual parasites and infills. To analyze the combinations and articulations of elements in the Recording Over pattern that produce the involute quality of space, it is necessary to understand the variations in the characteristics of these elements. At the next level of the hierarchy, the units of elemental space are the subassemblies of elements generated within a clearly framed space, grouped according to their characteristics based on their typology (i.e., parasites or infills), origin (e.g., organic or resurrected), or some other specific factor. Moving up the hierarchy, the subsystems coincide with the frames' identifiable configuration (Fig. 6.25), defining each of the three building blocks within the selected urban stripe: *Shallvare-Palazzine, Puna,* and *Gjykata.* Finally, the upper level of the hierarchy comprises the entire system structure. The Recording Over pattern is found in urban entities in which levels of hierarchy exist both in relation to

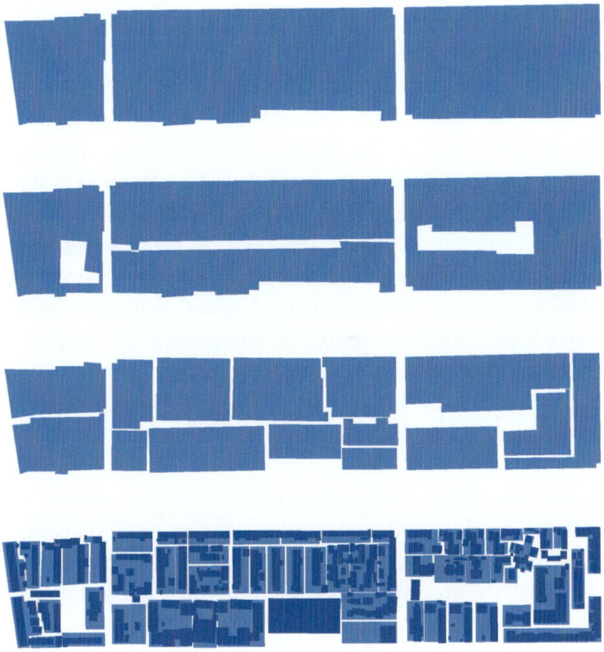

Fig. 6.25 Levels of hierarchy within the system; the whole and its parts manifest the same principles of irregularity, creating a self-affinitive phenomenon

the city (Figs. 5.29 and 5.30) and within the entities. This hierarchy derives from the mechanistic logic deployed by the centralized planning process, in which the city was conceived as standardized additional boxes.

To further understand the fractal properties of this pattern, it is important to study the behavior of the elements within the Recording Over system structure, or the internal dynamics that create that form. As already mentioned, the patterning process is set into motion by a motive. Here, this was the re-appropriation of public space, which triggered repetitive inward behavioral actions from scale to scale. This interfering pattern triggered the dynamics of the system structure from the bottom of the system. This settlement behavior at the bottom (Fig. 6.12) is reflected in the modality of wrapping at the macroscopic level as a recursive action, or the number of steps and sequence of actions, which acts to record over the frames or implode the space between them. These principles, once transformed into a shared experience or modality, become the basis for the creation of self-similar or self-affinitive physical forms through the levels of the system structure, self-regulating the system from the top. As such, they pervade the entire system structure and order the behavior generating this pattern. These internal systems of interrelationships constitute the dynamics that define the morphological expressions of the Recording Over pattern. Each of the abovementioned levels is guaranteed by the same motive: the re-appropriation of public space. The resulting self-similar or self-affinitive involute quality of space through structure, manifests the whole in parts.

The behavior of the elements and their articulation within the subsystems depends on some factors: the type of space-occupiers (e.g., parasites or infills); the type of subsystem space or typology of space within the frames; the functional typology (e.g., housing, commercial services, restaurants, or parking); the formal typology (e.g., low-rise villas or multistory buildings and their formal layout and position); and the origin of the Recording Over element and the type of layer in which it emerged (e.g., original organic, first superimposed, resurrected, or second superimposed; Fig. 5.28). Thus, redefining the involute quality of space from the fractal perspective, we can characterize it as a space re-appropriation process of the preexisting centrally designed emptiness through decomposing from the bottom or recomposing from the top of the system a series of self-similar or self-affinitive subassemblies, generated by individually arranged recursive agreements to parasite or fill in the space.

As already mentioned, closing the space, even in this pattern, was an echo of the Historical Organic pattern. Therefore, during the continuous and gradual densification process through wrapping by parasitic and infilling structures, spatial closing was driven by a desire to recursively reinforce and consolidate the embryonic characteristics of the involute quality of space (Fig. 5.33). Once again, the geometry of this spontaneous order has a specific meaning and is a direct result of urban improvisations emerging within a specific social and cultural construct. To understand this, we studied the anthropological principles as the main event that generated this new emerging reality. This provides insight into the human properties of this pattern, which may be used as an enzyme during the design of a generative process.

6.3.3 *Compressed New Organic*

The following two sections explore from the fractal perspective the emergence of the New Organic pattern in both its collapsed and inflated forms. The Collapsed New Organic pattern will be discussed first, with the aim of understanding the organizing principles of the system structure and the dynamics that generate the collapsed-in-on-itself quality of space. Figure 6.26 illustrates the repetitiveness of the basic elements, which create a self-affinitive essential quality at different levels of hierarchy.

The patterning process, in this case, was set into motion by the motive of isolation and amassing, which required densification as a repetitive auto-protective attitude or personal behavior, acting at the bottom. This behavior triggers the dynamics of the system structure and is reflected in the modality of compression as a shared experience at the top of the system. This modality of recursive compression, emerging from competing of individual behaviors, applied through scales regulates and orders the occupational steps and their related elements, such as the generative path and the parcels. Pervading the entire system, the modality self-regulates from the top

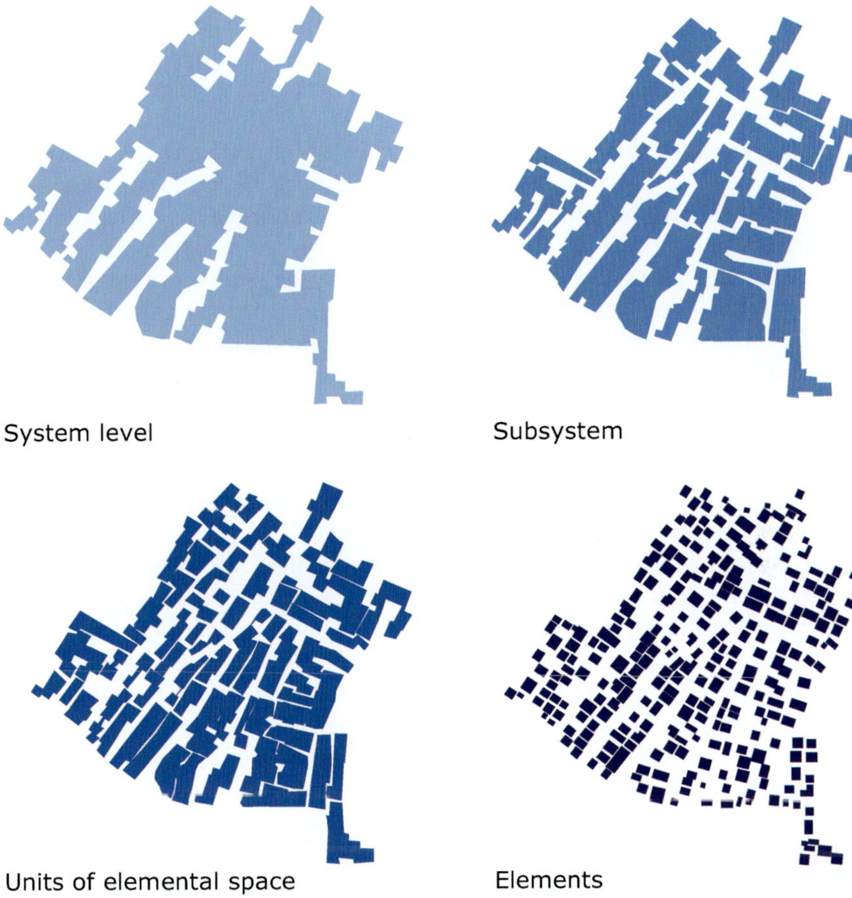

Fig. 6.26 Levels of hierarchy within the system. In this self-affinitive system structure, isolation and amassing are applied in the settlement process

a self-affinitive structure. These dynamic interrelationships are materialized in the morphological expressions of the Compressed New Organic.

At first glance, it is difficult to organize the elements of this pattern into legible motives; however, the smaller fragments exhibit an underlying structure and distinctive character that seem to obey a logic similar to that of the whole. At least four self-similar levels can be distinguished in which the collapsed-in-on-itself quality repeats according to similar principles. In most cases, the morphological weaving of the New Organic areas, among others, is influenced by the layout of the preexisting former agricultural service roads and land parcels. The four levels of hierarchy in the analyzed area, from individual elements to the system as a whole, are shown in Fig. 6.26.

At the first level are the elements of the system, including the houses, streets, dead-ends, tiny squares, plots, and fences. These are the fundamental physical components at the base of the pattern, the behavior of which is influenced by the motive materializing in the individual arrangements. These elements clash and approach a near-to-collapse condition (not necessarily physically, but also mentally through interfering patterns), generating the collapsed-in-on-itself quality of this pattern. At the next level of the hierarchy, the units of elemental space are formed by aggregations or subassemblies of elements (e.g., clusters of houses) around tertiary or lower-level internal paths. Next, the subsystems are represented by the freeform stripes resulting from the interactions between the units of elemental space and their respective access paths. The quality of amassing, transmitted from the lower levels, is reinforced at this level and embodied in a tight urban knitting; the competition between groups becomes evident here. The entire sample, the system, sits at the top of the system structure. At this level, the full morphological expression of the collapsed quality of space emerges from the interaction of the abovementioned self-similar subassemblies. This level can be considered the external form, containing the organizational structure and the full cycle of the internal dynamics.

In conclusion, we can say that the coherency of the motive, behavior, and modality, rooted in specific socio-historical and mythical circumstances, generates the self-similarity or self-affinitive collapsed-in-on-itself quality of space through structure, manifesting the parts in whole (Fig. 6.27). As such, this is not the geometry of a spontaneous uncoordinated order, often simply classified as informal; contrarily, it is the expression of a specific meaning, forged under the unbroken web of wholeness. The shift from merely considering these areas as informal to understanding the organizing principles and human properties underpinning this pattern represents a substantial step toward the deployment of this understanding in a generative process for a more natural city.

Redefining the collapsed-in-on-itself quality of space from the fractal perspective, we can characterize it as the quality of space forged under the combination of two main dynamics. The first is a densification or amassing behavior acting from the individual level at the bottom to the neighborhood level at the top, while the second is a compression modality acting from the top of the system to the lower levels, creating a cascade of self-similar subassemblies through the levels of which the collapsed-in-on-itself quality of space repeats with a similar logic.

6.3.4 Inflated New Organic

In this section, I explore from the fractal perspective the recursive irregularities characterizing the inflated quality of space and how they repeat geometrically through scales (Fig. 6.28). The internal dynamics of the system structure in this case are triggered by a dispersive behavior, as a repetitive and auto-protective attitude at the bottom of the system, stemming from the motive of isolation and loneliness. As mentioned in Sect. 6.2.3, these dynamics are regulated and ordered by the modality

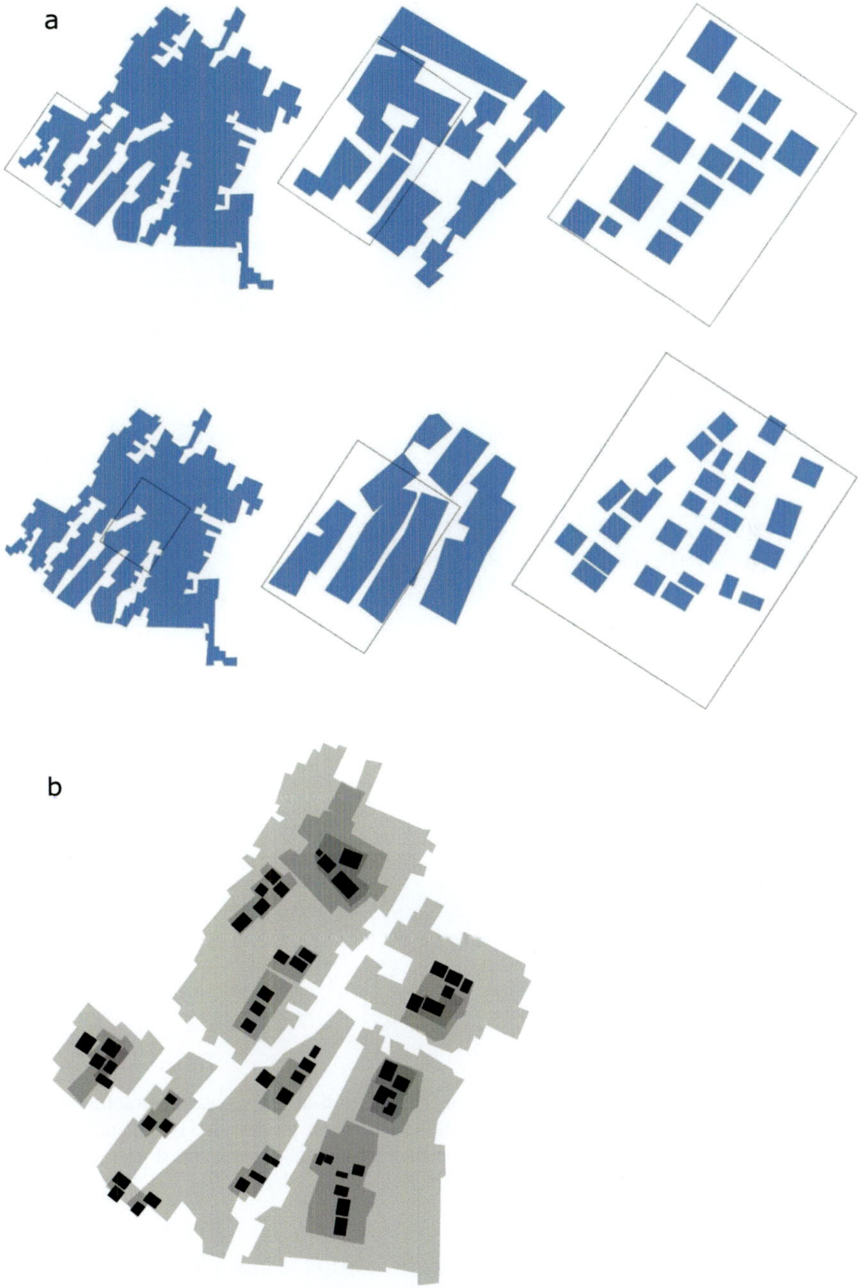

Fig. 6.27 **a** The whole in parts. From the system to its elements, a similar kind of irregularity is manifested in whole and in parts. Despite being on different scales, the parts of the system seem like continuations of each other. This pattern is characterized by self-affinitive principles that scale up and down to generate the collapsed quality of space. **b** The modality of recursive compression across scales, materialized in clashing areas as a self-affinitive principle that scales up and down to generate the collapsed-in-on-itself quality of space

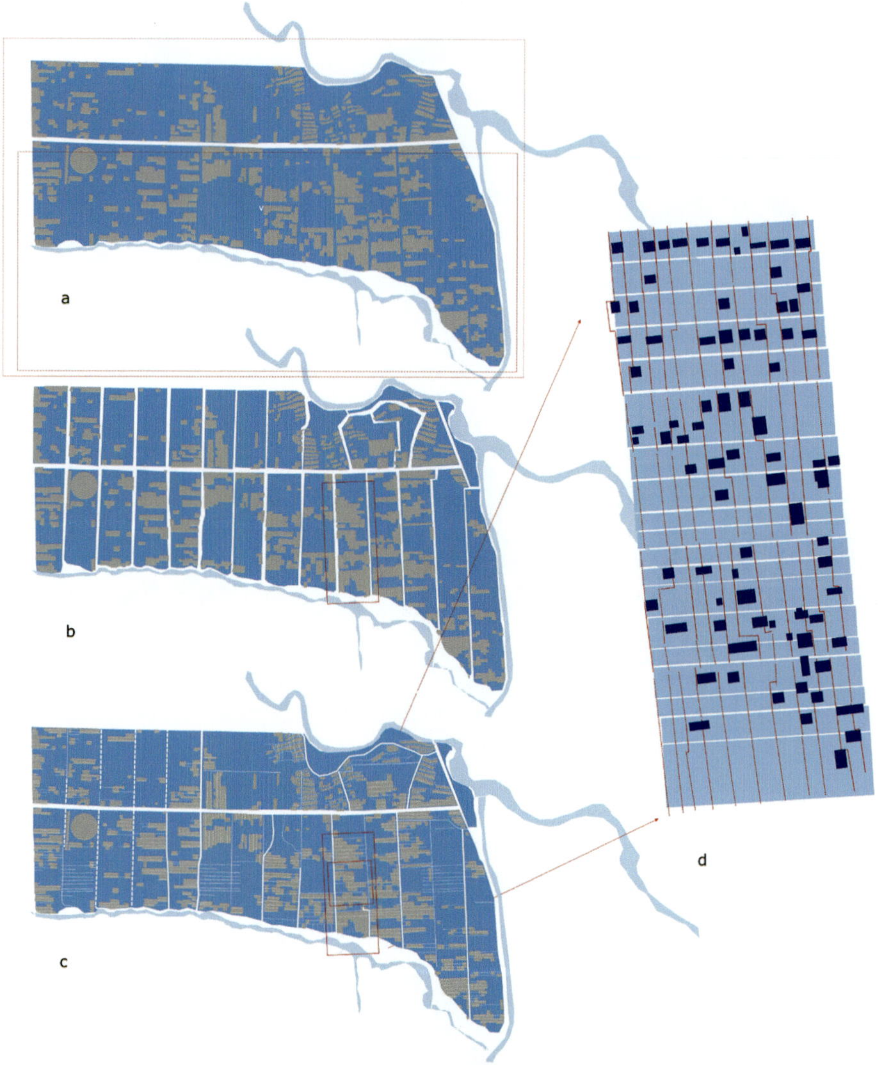

Fig. 6.28 Levels of hierarchy within the system. This is a self-affinitive system structure in which the motive of isolation in loneliness is applied in the settlement process. **a** System and subsystem levels (marked with red quadrats); **b** units of elemental space; **c** elements; **d** zoomed view of transversal stripes, houses, and perceptive lines (marked in red color)

of floating, which acts as a social norm from the top to the lower levels of the system. Thus, the settlement procedures and all elements of the system self-regulate in a self-similar logic according to this socially accepted norm that pervades the entire system and generates cyclical repetitions from the top. These dynamic interrelationships are materialized in the self-similar geometric logic evident in the morphological expressions of the Inflated New Organic.

On initial inspection, the area representing the Inflated New Organic appears to have a regular geometric structure. However, this is only its external appearance—the skeleton upon which the meaning of the inflated quality of space is woven. While the geometry can be regular or organic, there is an underlying structure at the smaller scale that respects the same logic of the whole and in which the specifics of this patterning model can be found. In the presented sample, the self-similarity of this pattern combines the almost perfect orthogonal modular geometric order of the preexisting rural infrastructure with the dynamics of the social improvisations explained through the categories of motive, behavior, and modality and rooted in the specific socio-historical background correlated within the web of wholeness. This interaction generates the irregularity of the basic elements corresponding to the quality of inflation that repeats through the four self-similar levels, as presented in Fig. 6.28.

The first of these levels is the elements, which sit at the bottom of the system, located within the transversal agricultural stripes (i.e., the smaller land subdivision). These stripes are based on the previous agricultural infrastructure, sitting between the third-level irrigation canals and perpendicular to the upper-level ones. Within these stripes, the individual housing plots are settled with specific housing modules that segment these stripes into smaller portions. The family may occupy one or more modules (segments) along the stripe, with each having a different use. In addition, smaller subdivisions and various functional activities, such as groves, cow houses, greenhouses, orchards, internal service roads, and water canals, may exist within each individual module. It is at this scale that the infinitesimal inflation logic applies. The 90-degree logic in this case is imposed by the geometry of the upper levels of the system.

At the second system level are the units of elemental space; that is, the longitudinal stripes between the second-level former irrigation and/or drainage canals and the cluster of houses along them. They are perpendicular to the previous agricultural stripes, reflecting a self-similar subdivision logic. Next, the subsystems consist of longitudinal stripes formed between the main irrigation and drainage canals, or the main agricultural roads, together with their respective clusters of houses. These subsystems are laid out in an almost rectangular way in relation to the previously analyzed ones. In some cases, these large-scale territorial connectors may be former service roads flanking riverbeds, embankments, or any other agricultural or territorial infrastructure, following a rationalized topographic logic, not necessarily in orthogonal relationship with other lines.

Finally, the system level is developed along two sides of a former main agricultural service road within the area and is limited on both sides (north and south) by two branches of the Lana River. All levels previously analyzed are contained within this system, and a subordinate self-similar logic of subdivision characterizes the whole and its parts (Fig. 6.29). This is considered the top of the system, which by itself is part of an agricultural grid and an integral subordinate part of it. Thus, the fractal

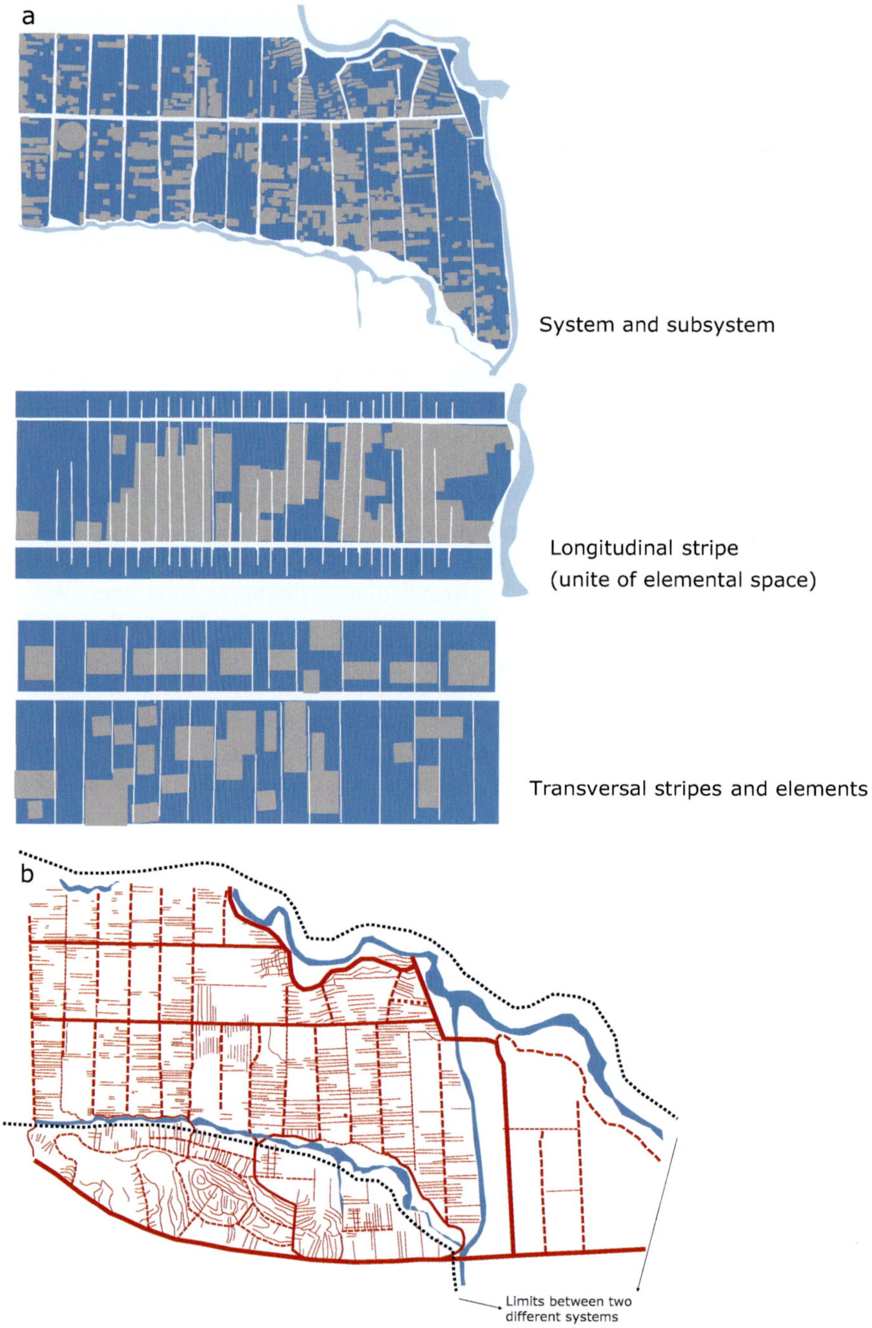

Fig. 6.29 The whole in parts. **a** From system to elements, a similar kind of irregularity is manifested in the whole and its parts (in land subdivision and house dispersion). The modality of recursive floating as a self-affinitive principle generates the inflated quality through scales. Despite being on different scales, the parts of the system seem to be continuations of each other. **b** Recursive irregularity of the generative path across scales as a self-affinitive principle that scales up and down

properties of this system in relation to the upper levels of the hierarchy refer to the larger agricultural territory.

At this point, we may ask how this model manages to preserve its main characteristics in the face of further developments and densifications within the same area. The key lies in the fact that the dwelling unit is conceived as an integral part of a self-similar territorial system or the focal point "grouping" around territorial elements. According to this logic, the consecutive subdivisions obey the same infinitesimal self-similar or self-affinitive logic, creating similar "entity containers." The replication of this relationship or mental pattern throughout the system means that these entity containers always contain units of territorial elements or uses grouped around the house (Fig. 5.52) at an infinitesimal distance. This is the DNA or human property of this pattern that creates the idea of a potentially inexhaustible and inflated space, regardless of how much space is actually available. In fact, a considerable portion of the transversal narrower stripes at the bottom of the system remains reserved for agricultural activities, which maintains the condition of distancing, understood as a particular kind of nearness that can be infinitely reduced without losing the qualities of the anthropological model.

In the analyzed case, the geometric network guides the settling structure, and at the same time is impregnated with specific human properties, as already described. A regular geometric structure deprived of a specific human meaning is just an empty shape. Here, we are reminded of the superficiality of the idea that a regular geometric structure by itself can improve living conditions in so-called informal settlements. In fact, it can be an alien intrusion if not well fitted in the web of wholeness, and if it fails to synchronize with the internal dynamics of the system.

Thus, in redefining the inflated quality of space from the fractal perspective, we can characterize it as the quality forged under the combination of two main dynamics: a dispersive behavior acting from the bottom and a floating modality acting from the top of the system, creating a cascade of self-similar subassemblies in which the inflation quality of space repeats in a similar logic across levels.

6.4 Complex Approach—Urban Patterns as the Human Capacity to Self-organize

The correlation effects and overlaps within the web of relationships that trigger the dynamics of the system structure and generate the relational order, explained above under the quantum and fractal approaches, perfectly fit within the complexity perspective. As mentioned, the analyzed patterns can be seen as the human capacity to self-organize or as relationships emerged within a self-organized collective whole.

Considering self-organization from the bottom as the main property of complex phenomena (Portugali 2000, p. vii), Mitchell's (2009, pp. 12–13) three principles of

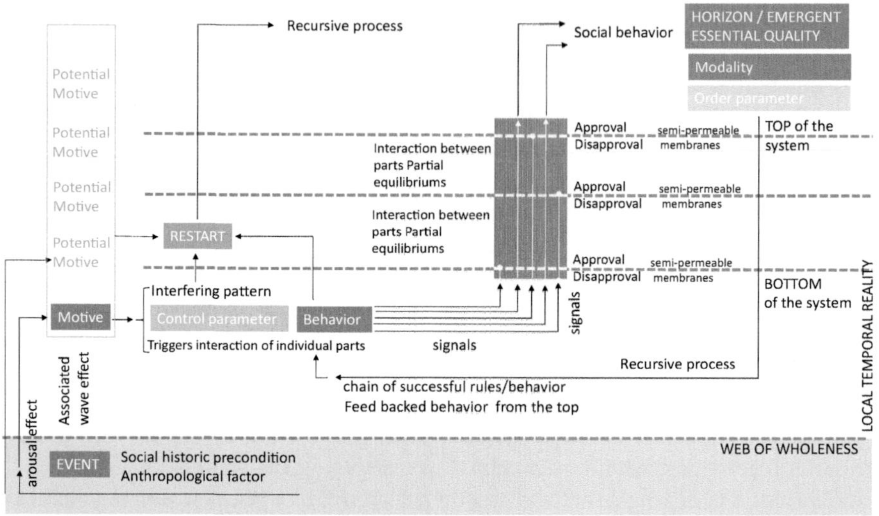

Fig. 6.30 The generation process of the new emergent reality across levels of the hierarchy. This figure shows the recursive nature of the process, the correlation factors, and the semi-permeable membranes of approval and disapproval that establish behaviors and rules

complex systems (outlined in Sect. 6.1.3), and Holland's theory of learning and adaptation in a CAS (2014, p. 35), here I give more details about interactions and adaptations through signaling, information-processing, and actions in a self-organized process. These kinds of complex systems are pervaded by a combination of top-down and bottom-up effects that generate interactions at different levels (Holland 2014, p. 29; Fig. 6.30). Figures 6.23, 6.25, 6.26, and 6.28 illustrate these system levels.

In a real settling situation, self-organization can be seen as emerging from both formal and informal governance arrangements, organized by the residents within the settlements, individually or collectively at various types of scales to enable the access to housing and/or other basic services. In these kinds of processes autonomous agents, or residents/settlers, engage in a "do it yourself development activity" by "mutually adapting and co-evolving" in complex adaptive systems/assemblage (Suhartini and Jones 2019, pp. 25, 29, 30, 229). As will be presented in the following paragraphs, collective intelligence, expressed in signals and processed into actions that affect the environment, raises in the system while agents share and exchange information and resources as a spontaneous response to achieve the goals. At the point that there is a shared understanding of the group members to achieve common goals, self-organization can be transformed in self-governance (Suhartini and Jones 2019, p. 30).

In the previous section, I analyzed the interactions between and within the four levels of the hierarchical system structure, each of which has a role in the process of generating new emergent realities. Regarding the hierarchical organization of CAS,

Holland (2014, pp. 64, 65) imagines the four levels as divided by a "succession of enclosing boundaries" or semi-permeable membranes that allow only those signals that fulfill the motive to pass (Fig. 6.30). Starting with the elements at the bottom of the system, these smaller agents interact based on "detectors" and "effectors" that translate the activity of other agents in the environment into "signals for internal processing" and "actions that affect the environment" (Holland 2014, p. 57). At this level, the behavior triggered by the motive materializes in individual arrangements and is transmitted as information through signals at the upper levels, which are then either confirmed or disconfirmed as a shared modality or norm. As an example of this, based on the different ways people close the space at the level of the individual house in the pursuit of introvert behavior (Fig. 6.5), only some of these actions will be confirmed as appropriate for guaranteeing the consistency of the motive. The subassemblies at the base of the system structure aggregate based on the affinity between social groups, starting from individual families through to clans and tribes, which serve as filters to confirm or disconfirm behaviors.

The next level, the units of elemental space, is formed by subassemblies of elements and acts as the first level of approval (from the defined groups or clans) that transfers individual behavior into a modality. At this level, an assessment is made of the usefulness of the different behaviors based on how well they help the agent to fulfill a motive. This is done by assigning a higher strength to a rule or behavior that leads directly to a result, and a lower strength to those that are rarely or never used. A rule or behavior starts becoming stronger if it is successful in strengthening its signals (based on competition between individuals and groups). Each passing behavior is strengthened by a signal activated from its precursor in the chain of behaviors, with positive feedback coming from the upper levels of the system. As an example, how space is closed at the level of the group fulfills the same need at the lower level of individual arrangements and the upper level of the subsystem. Unsuccessful or unshared behaviors do not pass this membrane (Holland 2014, p. 58) of confirmation. A typical example is shown in Fig. 6.8, where a series of secret rooms serve as a filter to confirm or disconfirm certain behaviors.

The subsystem level is dominated by interactions between units of elemental space or competition between groups based on clans or origin, to set the predominating imposing shared behavior (Fig. 6.6) before clashing or homogenizing with the entire system structure or the top of the system. Confusion or a lack of clarity at this level creates symptoms that affect the entire structure. This level thus acts as a kind of pre-filter, identifying the winning rules or behaviors, ready for them to be confirmed at the upper level as a social modality. This level reconfirms and reinforces the strength of a rule or behavior based on the "direct interactions with (1) the rules sending signals to it, and (2) the rules directly acting on its signals" (Holland 2014, p. 59). In other words, at this level, verification occurs regarding the success of the rules or behavior used at the lower levels (e.g., closing the space as a quality of concealing) in guaranteeing the same aim at the upper level. This involves confirmation that the concealing factor (rooted in the principles explained in Chaps. 4 and 5), as a standard and recurrent feature of the Historical Organic pattern, is respected at all levels of the hierarchy. Finally, based on competition and interaction, a rule may be improved or changed.

At the system level, the full cycle of the internal dynamics is manifested. This is the top of the system structure, where the rules are set—by confirming strong rules and disconfirming weak ones—and fed back into the system (Fig. 6.7). These rules order the entire system through a cyclical self-organization of the successful standards (rules or behaviors) generated from the system itself, acting like feedback loops. These loops involve a "large number of sequences" that "recirculate signals" and "offer control through positive and negative feedback" (Holland 2014, p. 71). At this point, the full chain of successful rules or behavior is displayed: its signal-processing effect has penetrated the entire structure, and feedback is coming from the top. This controls the behavior at the lower levels of the system, imposing order parameters from the top as a shared experience or modality such that more people follow the same rules. The replacement of weaker rules with new rules, to adopt the agents' rules at a higher level, is a "biased" (2014, p. 60) process, driven by archetypes based on the agents' previous experiences. According to Holland, "it has been observed that innovation in CAS is mostly a matter of combining well-known components in new ways" (2014, p. 60).

As mentioned, Portugali (2000, p. 58) argued that in the cognitive system of a person the wining order parameter of several configurations, resulting from competing of associative memories "enslaves the cognitive system." This is how the repertoire of motive stabilizes or sets its range during the competition between several memorized configurations, until some (that fulfill the agents' aim to close the space in a specific way triggered by the motive) receive positive feedback from the top (as shown in Figs. 6.5, 6.6 and 6.7; also Figs. 5.2 and 5.3). Thus, agents at the bottom adapt by "changing [their] signal-processing" rules or behavior to align with structural changes in the system's signal-processing networks (Holland 2014, p. 58). The rules or behavior that receive positive feedback are strengthened, becoming more successful. In this way, the repertoire of motives or archetypes, (Figs. 6.5, 6.6 and 6.7), is affirmed.

In the case of the Historical Organic pattern, the associative memories related to the motive of the right to visual privacy, as the associated wave factor of the original event, generated refractions and bifurcations as the winning configurations (Figs. 6.5, 6.6 and 6.7), or in other words, as the acting modality, or order parameter, emerging from a shared group experience. These building blocks gradually translated into anthropological principles, constituting the self-organization motive configuration for shaping the visceral quality of space (the essential quality of this pattern). This motive configuration was similar but not the same as the original event within the web of wholeness that generated and enacted the associative memories. This motive eventually became a group ability (macroscopic behavior) to reproduce the specific quality of space at different scales, including at the territorial and urban hierarchical levels. In this way, Eden(s) were reproduced that guaranteed independence from the external world through inwardly arranged enclosures, starting from the neighborhood level down to the tiniest unit of intimacy, which was impenetrable to the view of outsiders or up to the completely privatized space in which the behavior of enclosure lost its significance, and influenced the same behavior. This complex process reinforced and recycled the same mental pattern in a cycle of reproduction.

Refractions and bifurcations were the winning configurations not only because they could fulfill the specific meaning of privacy within the "Muslim right to visual privacy" (Kostov 2003, p. 63), but also because of the flexibility they offered in relation to the endless individual informal arrangements for modifications (e.g., plot subdivisions and the opening and closing of both interior and exterior space). This was also one of the winning criteria (rationality) underpinning the success of refraction and bifurcation in the Compressed New Organic pattern.

A similar reasoning can be followed for the other winning configurations in the analyzed patterns. For example, the associative memories related to the motive of the re-appropriation of public space (driven by organic land fragmentation) gave rise to the wrapping of space through parasitic and infilling structures as the winning motive configurations (Figs. 5.25, 5.26, and 6.11) for shaping the involute quality of space. In the case of the two versions of the New Organic pattern, collapsed-in-on-itself and inflated, the associative memories related to isolation expressed through amassing on the one hand and loneliness on the other generated, respectively, the compression and dispersion of space as the winning self-organization motive configurations (Figs. 5.38, 5.46 and 6.20).

This common path through the self-organization process demonstrates how the system is set in motion, how its smallest parts interact to produce macroscopic behavior that generates patterns and their essential quality at the upper level (i.e., visceral, involute, collapsed, or inflated), and how the system self-reproduces "circularly" and casually (Portugali 2000, pp. 57–58). This analytical path starts from conceptualizing urban patterns as a complex behavior (i.e., the self-organization of a large number of individuals with the aim of solving housing problems outside central control). The quantum and fractal approaches help to better understand the real phenomenon and the detection and distillation of the human properties behind these apparently abstract and indecipherable geometries. How this information can then be translated into input is the subject of Chap. 7.

The common path outlined above is only possible within the framework of an open system. Only open systems are suited to complexification. Within a specific social and cultural construct, even the analyzed urban patterns behave like open systems. As such, they are characterized by temporary equilibriums that, based on a social formula for the use of space or other shared or non-shared criteria, are potentially adaptable and correctible at any time. This kind of behavior can be seen as the propensity of the open system to adapt toward a never-reached order, or a potential condition of order aiming for self-regulation through learning to reach a higher level of complexity. Or as we called before in this section, a potential condition of self-governance, considered as a shared understanding of the group members to achieve common goals. This is an important human property that can be distilled from a better understanding of complex conditions and then transferred as an important principle for a generative design process.

For example, the visceral quality of space and its related formal configurations (i.e., avoiding, concealing, refracting, bifurcating, wrapping) underpin information exchange or group intelligence to guarantee a certain continuity related to the essential quality of space. From self-organized behavior emerges an ability of groups to self-regulate based on social units (e.g., clans, tribes, or place of origin) to reach a higher order of complexity and approach a new temporary equilibrium. Similarly, the involute, collapsed, and inflated essential qualities of space constitute the genetic code transmitted through information exchange during the settlement process. Through involvement in the common path of the self-organized process, the inward or condensing–dispersing individual behaviors become community stratagems that act as vehicles to shift the system toward new temporary equilibriums: the Recording Over, Compressed New Organic, and Inflated New Organic patterns. Importantly, to avoid obstructing the self-organized cyclical and interactive path, the system should not be closed or sterilized through linear thinking (e.g., zoning).

Notably, the Recording Over pattern represents a transition from a centrally controlled projection of space to a self-organized bottom-up and atomized one. This change from a degree of order to an almost entropic condition continuously searching to achieve a higher-level equilibrium is typified by the residential blocks built during the dictatorship, which were originally intended to be eternally and absolutely closed systems. Once the system opened, a more complex condition was created: the system declined in disorder and increased in entropy through the wrapping of the original system in parasitic and infilling structures. As a result of this complex behavior, these open-spaced blocks gradually transformed into involute organic blocks (Fig. 5.33). After the socio-political and cultural shift that marked the end of the era of dictatorship, the ownership patterns of the previously forcibly collectivized and unified space changed, and the unit of occupational structure through building moved from the large frames to the smallest possible cell that could be adapted or parasitized to redirect energy from previously established buildings and spaces. This is a central characteristic of complex systems. As Mitchell (2009, p. 4) observed, people "organize themselves, without the benefit of any central controller, into a collective whole that creates patterns, uses information … and evolves" through learning.

Even though the Recording Over pattern tends to reach a state of self-regulation, the current situation in these areas is mostly one of precarious equilibrium, sometimes even similar to the natural selection process. To positively influence the continuous search for higher-level equilibriums, we need to first understand how the system structure as a whole is generated, how the dynamics or behavior of the system tend to work, and how the system evolved and reached a certain degree of equilibrium. Then, we need to understand the internal resources of the system and its environment and, crucially, how the system structure can reach a higher degree of self-organization by borrowing energy from the system to potentially drive a "higher order complexity" and "differentiation" (Arida 2002, pp. 173–174). This complex condition requires the identification of a social formula for the use of space and the perceived consent process.

In this sense, the Recording Over pattern emerges only within an open system, maintaining or reaching temporary equilibrium based on information and energy exchanged within the system. Recording Over as a pattern stops existing, or gradually atrophies to death, once the system is closed and disconnected from the environment that generated it. However, how these blocks self-organize and potentially self-regulate as open systems remain challenging to apply. As such, it would be valuable to create a model through which to study this behavior.

The same is true for the Compressed New Organic. Surveys conducted in these areas provide evidence about some of the stages in the self-organized settlement process. First, the existence of a certain degree of natural order contains traces of initial urbanization that guide the self-organized settlement process. Second, a gradual decline in disorder can be expected as more constructions appear in the same territory (an increased level of entropy) and because the level of communication and information exchange between entities is initially quite low (signaling a problem with the process). Third, if the system remains open, further developments may help to reach certain dynamic equilibria; the need to communicate individually and share experiences in groups increases, and with it, the ability to governance arrangements to achieve social and physical self-organization, which becomes the energy and dynamics of the system. Fourth, the system may gradually reach a higher level of order and partial equilibria or relative consolidation. Fifth, this precarious equilibrium may be upset by external interventions (e.g., enclosures of the system or the arrival of new groups of people) or internal problems (e.g., atrophying of activities and the malfunctioning of community dynamics and communication due to a lack of trust and social tensions in response to heterogeneous historical and anthropological backgrounds). This may create conditions for a gradual decline in disorder, with the process of governance arrangements reflected in self-organization and self-regulation restarting a new cycle until equilibrium is established.

While there are factors that lead to a decline in disorder, there are also factors that support co-evolution and help to move toward a new equilibrium. For example, external factors, such as government legislation and decisions for housing legalization (Law N. 9482, 03.04.2006) or projects to support the urban upgrading of these areas through service and infrastructure improvements, help to establish a new equilibrium and potentially a higher self-organization stage (Fig. 5.43). Although these kinds of interventions may be initiated and supervised by a central authority (e.g., NGO or government), it is important to implement them in coordination with local Community-Based Organizations (CBO) to match the social reality. The first official pilot examples that followed this model of neighborhood upgrading in Tirana (labeled informal) during the 1990s, were initiated by Co-plan, a local NGO (Social-economic report 1997), and catalyzed through CBOs or other external and internal factors (The World Bank Report 1998, pp. 4–6, 11–13, 33, 37). Negotiation within this clashing reality without the engagement of external leverage often became a barrier, reinforced by the lack of trust in central authorities. For this, the methodology of such projects should guarantee that the system dynamics preserve the characteristics of an open system within a co-evolutionary scenario where formal and informal institutions evolve, learn and adapt in the process; that dynamics are not imposed; and that the system borrows energy from its own environment in terms of internal human and physical resources (e.g., labor force, existing infrastructure, building materials, land resources, and local economic activities).

Creating conditions for the open system to perform is the only way to move from self-organization to self-regulation and to avoid entropy (Arida 2002, p. 141). For this, we must guarantee a healthy relationship between the various systems existing within these areas (e.g., systems of production, commerce and the exchange of goods, and local public foci), regardless of whether these are formal or informal. For example, a great deal of an area's informal production activities (e.g., building materials, furniture, and agricultural products) are used almost exclusively within that same area. Thus, rather than closing down these activities because they are informal, the government should support their further development and improvement to guarantee the auto-sufficiency of the internal economic system. Sporadic and tentative efforts by the government to regulate (or close down) the system by imposing a top-down and mechanistic order without understanding how the system works does little more than deteriorate already precarious equilibriums.

6.5 Conclusions

This chapter discussed the understanding of cities and their respective urban patterns as emergent phenomena, considering the theoretical lenses of quantum, fractal and complexity approaches. This understanding brought more clarity about the internal word of interrelationships and the conceptualization of the urban patterns as complex systems emerging from quantum correlations, fractal relational behavior, and interactions in a complex self-organized cyclical process. Therefore, an observation and analytical path to approach urban patterns as emergent realities start to become clearer, as outlined in the following paragraph.

The essential quality of patterns was shown to emerge from the arousal effect of relationships set in motion by a motive acting in the system as a control parameter, which triggers the behavior and interaction of the smaller parts, the agents at the bottom. This gives rise to a modality as an order parameter (e.g., social norms and collective intelligence), generated at the top of the system or at the macroscopic level that enslaves the behavior of the smaller parts. In this way, the order parameter is cyclically reproduced (Kelso et al. 2016, pp. 44–46) and enters into a recursive self-similar process. The internal structure and dynamics are precarious equilibria, subject to repetitive processes that start from the bottom and are fed back cyclically from the macroscopic behavior of the entire system to reinforce and restart a new process. These dynamic equilibria are representative of a complex adaptive system or assemblage resulting from a range of formal/informal arrangements and actions undertaken by residents/settlers that stay at the base of the transformations. The intelligence that raises through levels of this complex adaptive system tends to transform the self-organized process starting from the bottom in a self-governance process, which implies a larger shared understanding about the mutual goals, or increased awareness at the upper level of the system. The generative process underlying each of the analyzed patterns is summarized below.

The Historical Organic (Fig. 5.21) is a self-organized settlement process in which recognizable patterns expressing the visceral quality of space emerge from a recursive process motivated by the right to visual privacy acting as a wave effect, which is reflected in the introvert behavior and the modality of refracting the straight lines. The visceral quality of space and its morphological expressions generated by these dynamic interrelations based on group affiliation or family ties is repeated or interlocked in a self-similar or self-affinitive way through four levels of hierarchy, at the same time within and independent from the external world. These levels acting as filters allow to pass only behaviors/actions coherent with the motive.

Recording Over (Fig. 5.33) is a self-organized process emerging in the preexisting centrally designed urban units that gradually transform both the frames defining the space and the space itself. A self-organized mass of individuals driven by an inward behavior coordinated at the smallest level and motivated by the re-appropriation of public space generates the modality of wrapping the space and its buildings from within and without. These dynamic interrelations generate recognizable patterns expressing an involute quality of space that is repeated in a self-similar or self-affinitive structure through four hierarchical levels. These levels filter behaviors/actions which are consistent with the motive. Recording Over is the underlying form that gives a specific meaning to the abstract and universal frames on which this pattern builds.

The New Organic (Fig. 5.53) is a self-organized settlement process in which recognizable patterns of individual houses grouped in the urban and rural space acquire mass through densification or dispersion behaviors. From this condition of isolating (acting as a wave effect) in specific areas to avoid authorities emerges two variations on the quality of space: a collapsed-in-on-itself quality of space based on the modality of compression, and an inflated quality of space based on the modality of floating in rural space. This recursive process generates a self-similarity and self-affinity of the collapsed-in-on-itself or inflated quality of space through four levels of hierarchy, acting as filters that confirm successful behaviors in relation to achieving the motive. Collapsed and inflated are specific group devices directly related to the need to ensure the longevity of the settlements that carry this quality.

In this sense, the complexity of the Historical Organic, Recording Over, and New Organic patterns can also be expressed as a number of steps or sequences of actions contained within these patterns (or in algorithmic terms). This matter is taken up in the next chapter where the main challenge is to outline a methodology for investigating the emergent realities and their related properties, which is the main purpose of the book.

References

Arida A (2002) Quantum city. Routledge

Batty M, Longley P (1994) Fractal cities—a geometry of form and function. Academic Press, London, San Diego, New York, Boston, Sydney, Tokyo, Toronto

Co-plan—Center for Habitat Development (1997) Social-economic updating for Breglumasi pilot site, Co-plan documents

Holland John H (2014) Complexity—a very short introduction. Oxford University Press

Kelso JAS, Stolk E, Portugali J (2016) Self-organization and design as a complementary pair (Chapter 3). Complexity, cognition, urban planning and design. Springer Proceedings in Complexity

Kostov S (2003) The city shaped—urban patterns and meanings through history. Bulfinch Press AOL Time Warner Book Group, Boston, New York, London, Fourth printing

Mitchell M (2009) Complexity—a guided tour. Oxford University Press

Portugali J (2000) Self-organization and the city. Springer, Berlin, Heidelberg, GmbH

Suhartini N, Jones P (2019) Urban governance and informal settlements—lessons from the City of Jayapura, Indonesia. Springer, The Urban Book Series

The World Bank (1998) Report No: 17694-ALB. Project appraisal document. Infrastructure sector, Europe and Central Asia Regional Office. Document of The World Bank

Thomo P, Muka A, Zarshati F, Martini GJ (2004) Vendbanime dhe banesa popullore shqiptare. Akademia e Shkencave të Shqipërisë, Insituti i Kulturës Popullore, Toena, Tiranë

Zohar D, Marshall I (1994) The quantum society—mind, physics, and a new social vision. Quill William Morrow, New York

Legislation

Babani private photo collection. Tirana 1994

Law N. 9482 dated 03.04.2006 "On legalization, urbanization and integration of illegal constructions"

Chapter 7
Methodology for Holistic Understanding of the Urban Patterns

Abstract The aim of this chapter is to outline a methodology for a holistic understanding of the urban patterns that based on empirical observations and analysis, aims at organizing that information into a scientific model. The presented methodology consists of three main steps: The first one is related to the conception of the reality as a larger whole (quantum) that contains an underlying order made of relationships and hierarchy (fractal), and is self-organized as a collective whole based on individual/group interactions (complex). The keyword for this step is holistic reality. The second step is related to the application of the above lenses to unpack the essential qualities that emerged as interrelationships generated by overlaps of past and present events (quantum); as recursive irregularity of elements repeated in a certain order (Fractal); and as cyclically reproduced individual/group interactions (Complex). The keyword for this step is unpacking interrelationships through analytical categories. In the third step, the reality conceptualized from a holistic and relational perspective is translated into information to generate a model for better observation of the reality, or for design purposes. Sequential steps for depicting the recursive irregularities characterizing patterns' essential qualities and translating those qualities into information are described.

Keywords Methodological steps · Holistic reality · Correlational · Recursive irregularity analytical categories · Information

In Chaps. 3 and 6, I argued for more suitable theoretical lenses and discussed how they could be applied to penetrate the internal world of the generative process. For each pattern, I attempted to uncover the invisible filaments or unbroken network of relationships behind their emergent qualities in the local temporal reality. I then analyzed the structure of the repetitive patterns, which contain both a range of formal variations and a consistency in their generative logic. Finally, I explained how this emerges from the interactions within the collective whole, which I describe as a complex behavior. In relation to that, an important issue I present in this chapter is how to translate this complex behavior into operational information, which can be externalized in terms of urban schemes embedding morphological properties like geometry, orientation, roads structure, etc.

S. Dhamo, *Understanding Emergent Urbanism*, The Urban Book Series,
https://doi.org/10.1007/978-3-030-82731-1_7

The aim of this chapter, therefore, is to outline a methodology for a holistic understanding of the urban patterns that based on empirical observations and analysis, aims at organizing that information into a scientific model. We use the notion of the scientific model in order to emphasize the understanding of urban patterns rather than simply enabling the way the patterns can be predicted. Therefore, the main scope of the model is the identification, extrapolation, and representation of site-specific urban qualities (as shown in this chapter and in Appendixes A and B). Due to the large number of diverse factors interacting here, one challenge is to express this complex behavior in quantitative and qualitative terms; that is, to measure the complexity (Mitchell 2009, p. 111). Thus, before outlining the methodology, I summarize some helpful methods for measuring complexity.

7.1 Measuring Patterns as Complex Behavior

In Chap. 6, I argued that the analyzed urban patterns emerge from a complex behavior. They fulfill the three criteria set by Mitchell (2009, pp. 12–13); that is, "networks of individual components," "signaling and information" exchange, and increased "chances of survival … through learning." As we saw, during the settlement process, people trying to adapt to changing situations enter into relationships and produce a collective whole, which is a precarious equilibrium. This collective whole and the resulting quality of space contain all of the information related to the generation of this quality. However, before trying to extract information from these qualities, it is worthwhile to examine some methods of measuring these qualities as complex systems. Mitchell (2009, pp. 13, 94–95, 111) presented some valuable methods for "measuring … complexity," including "complexity as algorithmic information content" (2009, pp. 98–100), "statistical complexity" (2009, p. 102), and "complexity as degree of hierarchy" (2009, pp. 109–111).

It is difficult to include complex systems in a "single measurement scale" because of the many interrelating dimensions (Mitchell 2009, p. 111). For this reason, there are several methods for measuring complexity. Complexity as algorithmic information content proposes that the "complexity of an object is the size of the shortest computer program that could generate a complete description of the object" (Kolmogorov, Chaitin, Solomonoff in Mitchell 2009, p. 98). In this case, a string of information can be quite simple with "low algorithmic information content" (e.g., ABABABAB) or "random" with "no discernible overall pattern to it" (e.g., ATCGGTACT; Mitchell 2009, p. 98). Other cases may fall somewhere in between, with correlations among different parts or combinations with some randomness. Therefore, we can speak of "effective complexity," defined as the information content of a string's set of regularities (Mitchell 2009, p. 98). For this reason, we start by describing the regularities and how they combine with randomness. Gell-Mann (in Mitchell 2009, p. 98) proposed that effective complexity assumes that "any given entity is composed of a combination of regularity and randomness." Referring to the above examples, the first string (ABABABAB) has a simple regularity with a "predictable structure" and a "low effective complexity." Conversely, the second string (ATCGGTACT) has no regularities; therefore, "no information [is] needed to describe its regularities." In this case, "while the algorithmic information content of the string itself is maximal ... its effective complexity—is zero." Thus, "very ordered and very random entities have low effective complexity" (Mitchell 2009, p. 99). These concepts will be applied to translating the essential quality of space into information and conceiving the measurement and data-elaboration process (Tables A.1–A.3).

Statistical complexity (Crutchfield and Young in Mitchell 2009, p. 102) "measures the minimum ... information about the past behavior of a system that is needed to optimally predict the statistical behavior of the system in the future." Thus, based on observations of the system, we can create a model in which "behavior is statistically indistinguishable from the behavior of the system itself" (Mitchell 2009, p. 102). According to this method, the system is considered a "message source," and its behavior is quantified as discrete "messages" (Shannon in Mitchell 2009, p. 102). This model permits the inclusion of random choices; therefore, string types such as ATCGGTACT could be included (Mitchell 2009, p. 102).

Complexity as a degree of hierarchy implies "the complex system [is] composed of subsystems that, in turn, have their own subsystems, and so on" (Simon in Mitchell 2009, p. 109). Fractal self-similar patterns are an example of this. According to this method, the most "common attributes of complex systems are hierarchy and near-decomposability." The latter implies that "there are many more strong interactions within a subsystem than between systems" (Mitchell 2009, pp. 109–110). This is supported by pattern analysis, which shows more interactions between the elements composing the units of elemental space than between the units themselves. As illustrated by the scheme in Fig. 6.30, some individual behaviors are disapproved from passing to the upper levels of the system. Another example can be seen in the frames of the Recording Over pattern. Once built, they cut portions of the city off from the rest, preventing organic (normal) interaction. However, as soon as the system opened, these frames were recorded over, wrapped with parasitic and infilling structures growing out of the thriving metabolism occurring within them.

7.2 Methodological Steps: From Observation and Analysis to Essential Quality and Information

Understanding specific urban phenomena starts with a thorough investigation of the complexity of urban life and human behavior in the anthropological and historical context. To this end, I focused on the spatial structure, underpinned by the individual behaviors and their interactions that create the essential quality of the patterns as macroscopic emergence of human behavior. In line with this logic, I now outline step by step a methodology (Fig. 7.1) for identifying and observing the emergent realities and their essential qualities, which can then be translated into a set of regularities and informative principles for use in the design of a generative process, or as input

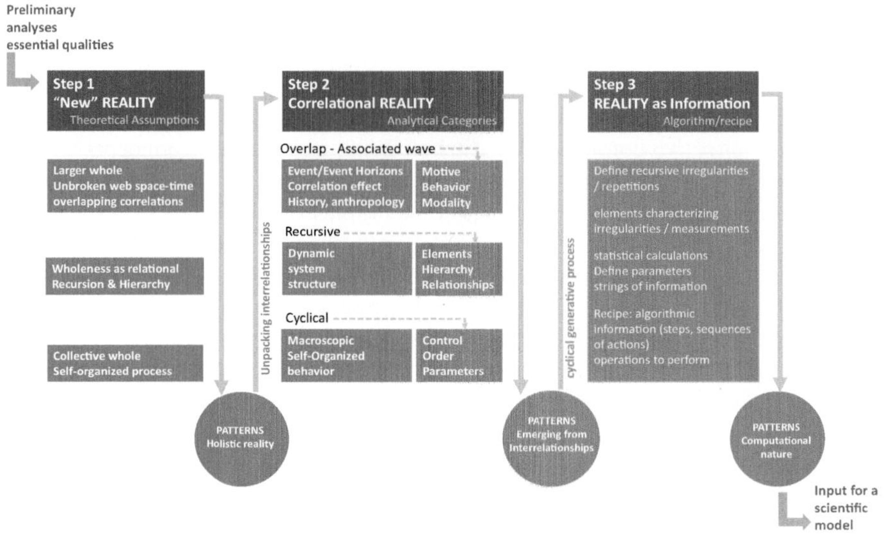

Fig. 7.1 Summary of the methodology

for a scientific model. As shown further in this chapter and in Appendix A, the first sample from the Historical Organic pattern is used here as an example.

Step 1: Main Theoretical Assumptions

As the first step, it is necessary to highlight a set of mental constructs, deriving from quantum, fractal, and complexity theories, which together support the main theoretical lenses to be applied in this methodology:

a. Reality is what we see, but also what we know and depends on which paradigm we use to observe the phenomena.
b. Reality is a container of infinite potentialities, interactions, and "tendencies to occur" (Arida 2002, pp. 55, 56).
c. The unbroken web of wholeness arises only through overlapping relationships transcending time and space—creating the "larger whole"; apparently separate things are aspects of a "larger whole" where things enter in correlation (Zohar and Marshall 1994, pp. 54, 58–60, 62).
d. A new emergent reality is not the sum of its parts because interrelationships are involved (Zohar and Marshall 1994, p. 62).
e. The territoriality of an event, or the associated event horizon, is more than spatial systems (because of the wave aspects). It is thus necessary to look beyond local temporal realities through the unbroken web of wholeness (Arida 2002, p. 150).
f. What appears superficially chaotic contains an underlying order based on relationships and hierarchy (Batty and Longley 1994, p. v).

g. "Self-organization" (Holland 2014, p. 31; Portugali 2000, p. vii) is a macroscopic behavior or "collective whole" that arises through interaction and information exchange (Mitchell 2009, p. 4). Other authors cited in the book bring additional meanings to this concept in the urban field, such as Dovey (2012) and Jones (2019) who speak about adaptation and self-organization at the base of the system's emergence; Marshall (2009) about city complexity and emergent urbanism; Kamalipour and Peimani (2019) about the micro-scale process of self-organization; Silva (2018) about self-organization and co-evolution through tactical actions. Self-organization is the basis of the complex behavior that produces structure and patterns, and also gives rise to intelligence.

The notion of the "quantum city" (Arida 2002) helped to introduce in the field of urban analysis a set of immaterial aspects under the concepts of the associated wave factor and interfering pattern, concretized in the specific analysis through correlation factors (i.e., motive, behavior, and modality), which were the main triggers for the conception of the essential qualities of each pattern as well as their continuous transformation. As such, introducing a quantum approach in urban analysis enabled a reconnection of the material and non-material aspects and gave meaning and awareness to urban space. The pattern analyses carried out in the previous chapters support the fact that shaping a space is not only a physical issue; before being physical structures, the essential qualities—visceral, involute, compressed, or inflated—were mental constructs or imprints. These qualities manifest the condensation of memory and meaning by which users saturate and impregnate the human space as part of their shared identification experience through generations, which transcends the limits of space–time. In this way, humans transform qualities of space into a potential cognitive information load, transferable as a non-material aspect outside their original environment and time in the form of an associated wave factor. This is how events (e.g., the essential quality of a pattern transformed in cognitive space) emit waves and create new or different but similar event horizons within the web of wholeness. In this sense, dwelling behavior and the way we exist (historically, in the present, past, and future) are powerful events that create housing horizons (i.e., as materialized in temporary physical borders). Efforts have been made to identify the limits at which the physical and mental meet; for example, the physical limit of a dwelling has been considered the event horizon of a dwelling behavior (i.e., the event creating it), while the dwelling is viewed as the main event creating the specific urban patterning process.

Understanding the city as a "field of information" (Arida 2002, p. 160) provides insight into how the cognitive space becomes material space and vice versa. It is clear that we cannot neglect this non-material factor, which has a primary influence on the generation of interactions and the creation of the physical environment with specific human properties. Because of these factors, there are similarities and differences between the analyzed patterns; when two or more waves overlap, the interference wave is different from the original one. By looking at the material and non-material aspects simultaneously, it is clear why the three analyzed patterns have some similarities, albeit transformed, despite arising from different space–time realities.

The methodological steps also account for the idea that the interrelationships that generate the essential quality of space embody some organizational principles that imply recursion of irregularities in a self-similar or self-affinitive system structure. Most importantly, this kind of organization, which can also be seen as part of the system's complexity, is how we investigated the irregularity of the natural city. Explaining and accepting this irregularity as a more complex form of order, removes the stigma of irregularity which is a factor hindering the acceptance of the natural city.

With this clearer understanding of reality as relationships that generate new emergent realities from overlapping relationships of associated wave factors, and as a self-organized collective whole that produces structure and patterns, we can proceed to Step 2.

Step 2: Holistic Observation; The Main Analytical Categories

a. Quantum approach: Explore history, culture, society, and anthropological background as part of the filaments weaving together the web of wholeness.
b. Explore the essential qualities of the urban patterns as emergent realities raised within the web of wholeness. Observe and analyze their morphological expressions in the local temporal reality as horizons generated by potential overlaps of past and present events.
c. Understand the arousal effect of these interrelationships and define motive, behavior, and modality as the correlation factors in the web of wholeness (space–time) that generate the essential quality of the new emergent reality.
d. Fractal approach: Observe and analyze the essential quality as a system structure or wholeness, generated by relationships (dynamics) between elements (structure, statics), creating the invisible structure or form. Link this with the correlation factors of motive, behavior, and modality and understand how these set into motion the elements of the system structure.
e. Identify the organizing principles; that is, the hierarchy of the self-similar and affinitive system structure, where the whole is reflected in parts. Understand the levels of the system (from the bottom to the top).
f. Identify the dynamics of the system; that is, the recursive irregularities through levels. Link these with the correlation factors.
g. Complexity approach: Observe and synthesize the essential quality of pattern as a macroscopic self-organized complex behavior, or as interrelationships created within a collective whole. Understand the cyclical self-organized generative process through the categories of the control parameter and order parameter, starting from individual behavior at the bottom of the system to the social modality being fed back from the top. Understand the interactions and competition between parts through signaling processing and actions. Link to, and verify with, conclusions from the previous steps.
h. Give human meaning and awareness to the generative process. For this, match and correlate the main analytical categories (i.e., event–event horizons, the dynamics and hierarchy of the system structure, the control parameters, and the order parameters) that create the cyclical generative process with the correlation

Table 7.1 Analytical categories for a holistic approach

Quantum Correlation effect	Fractal Dynamic system structure	Complex Cyclical self-organized behavior
Event/Event horizons Correlation factors: Motive (Associated wave) Behavior (Interfering pattern) Modality	Elements of the system Subassemblies Levels of hierarchy Behavior/dynamic of the system Recursions (Modality) Sequences of irregularity Archetypes	Bottom of the system Control parameter (Behavior) Order parameter (Modality) Signaling Processing Actions Permissive membranes Confirm/disconfirm Equilibrium/threshold Top of the system

factors (i.e., motive, behavior, and modality) that embody the abovementioned analytical categories. Observe their effect at each level of the system structure and make them evident in the story of the new emergent reality or new event.

i. Translate the cyclical (recursive) generative process into a number of steps and sequences of actions; that is, into algorithmic information in the form of a recipe (see Step 3).
j. Produce drawings and diagrams, for revision and improvement at each step—that reflect the understanding of the observed phenomena and their related morphological aspects, at each stage.

The reasoning followed so far is illustrative of reality's dependence on the paradigms we use to observe it and on what we already know about it. As such, I have sought to reframe reality as a transference of information, stored as human experience, from the field of the unconscious to that of consciousness. This includes things that are not visible at first sight or may appear completely different.

The methodology for observing and analyzing includes steps that seek to explain factors that normally remain hidden within the unconscious. This includes the correlation factors of the macroscopic behavior aroused in the web of wholeness that generates pattern formation; the dynamics of the system structure; the materialization of these dynamics in human behavior and competing interactions until they transform into a modality; the collective or group intelligence and awareness raised from information exchanged within the system to reach a higher level of complexity and a new precarious equilibrium; and the way the system evolves through learning. Achieving this necessitated the inclusion of additional theoretical lenses. For the main analytical categories see Table 7.1.

Step 3: Building the Model: Algorithmic Information

The steps presented above contain the theoretical foundations and methodology for identifying the essential qualities and holistic properties of the patterns. Next, the process for defining the algorithmic information is outlined, preceded by how the behavior of the system as a field of information can be quantified and statistically calculated. This will lead to data that can serve as input for a potential model. Note that these steps require the creation of detailed drawings and/or diagrams (Fig. 7.2).

a. Identify the recursive irregularities that characterize a range of forms relative to the essential quality of space and how they repeat through a number of steps and sequences of actions (identified from the previous step) through the hierarchical system (e.g., the repetition of deviation that characterizes the visceral quality of space as a recursive irregularity through levels of hierarchy).
b. Define specific elements that characterize those irregularities (e.g., deviation is defined by lines, refractions, bifurcations, angles, and rotations).
c. Measure and register those elements against factors, such as segmental length, angle (degrees), and direction of rotation. Organize the data in a preliminary table.
d. Calculate the relative frequencies or index numbers of the measured elements and perform any other statistical calculations that may be required for use as input in the modeling process. Define ranges and other elements as the basis for the distribution of values during the calculation process. (See Appendix A for more information on this.)
e. Define the parameters; that is, the quantified features of the recursive irregularities (e.g., the relative frequencies of the angle of refraction [Ar] for the ranges 1°–10°, 10°–30°, 30°–60°, and 60°–90° are 45%, 0%, 33%, and 22%, respectively). These parameters express the recursive behavior of deviation for one of the lines according to the predefined ranges of Ar. Recall here that deviation is the main recursive irregularity characterizing the visceral quality of space.
f. Write strings of information (e.g., the range of Ar for 1°–10° is 68% at the first level of the hierarchy, 53% at the second-highest level, and 6 and 7% at the lower levels), as well as the sequences and alternation sequences characterizing the direction of rotation, as shown in Tables (Appendix A, Tables A.1–A.3).
g. Define and describe the operations to be performed as a recipe that contains the algorithmic information and statistical calculations (see Appendix B, Sect. B.1).
h. Consider that it is an open-ended generation/design solution.

The first sample (area 1.1 in Figs. 5.1 and 7.2) from the Historical Organic pattern is used here as an example to illustrate the empirical observations and measurement process.

Pattern: Historical Organic, Representative area 1.1
Essential quality of space: visceral.
Recursive irregularity 1: deviating in a recursive logic (identify specific elements that indicate this behavior);
Reminder: the urban condition relative to the visceral quality of space.
A repetitive process of path deviation, motivated by the right to visual privacy and with the aim of isolation and concealment from the outer space through an inward aggregation. This is visible in the recursive logic of

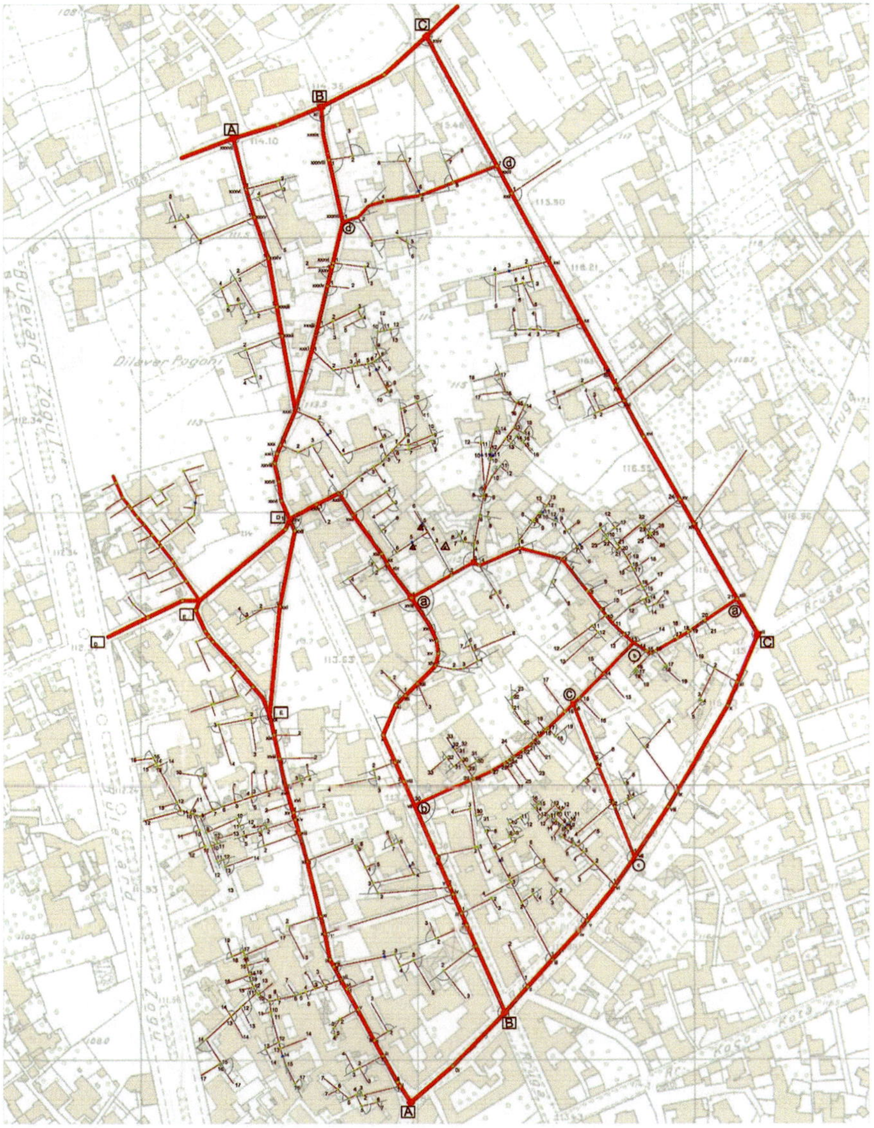

Fig. 7.2 Sample area 1.1: the basis for the observation and measurement of the elements and factors that define the recursive deviation

refraction and bifurcation and the related subdivision of space during the settlement process, which resulted in a self-similar or self-affinitive visceral quality of space across levels.

We start by identifying in diagrams the recursive irregularities of deviation, followed by the specific elements that indicate this behavior, which will be observed and measured against factors. Table 7.2 presents a basic matrix containing, in their respective columns, the elements defining the recursive irregularities (e.g., the nodes, lines, and angles), the factors indicating the dynamics or behavior of the elements (e.g., number of entities and length), and the measurements used for each of the above factors and the calculated parameters (e.g., relative frequencies and index numbers). This simple matrix is the basis for a more complex one (as shown in Appendix A, Table A.4), which combines all relevant data from the previous step with the statistical operations and highlights of the key parameters; that is the ones that synthesize the essential qualities. Note that the list presented in Table 7.2 is illustrative rather than exhaustive. Different factors or alternative measurement methodologies could have been followed, and further statistical elaboration and interpretation is needed.

Table 7.2 Recursive irregularity of deviation characterizing the visceral quality of the Historical Organic pattern: elements and factors to be measured

Elements defining the recursive irregularities	**Factors** indicating the dynamics or behavior of the elements	**Measurements** for each of the factors; relative frequencies and index numbers
Nodes	Number of nodes along the line (**Nn**)	**Nn**: an absolute number of refraction (R) and bifurcation (B) nodes **Nl**: Index number of nodes per length **Nlf**: Relative frequency of Nl
Lines/segments	Straight-line segmental length (**Ls**)	**Ls**: Directly measured distance **Lsf**: Relative frequency of Ls **Lsi**: Segmental length index number **Lsfi**: Relative frequency of Lsi
Angles	Angle of refraction (**Ar**)	**Ar**: Directly measured angle **Arf**: Relative frequency of Ar
	Rotation of refraction (**Rr**)	**Rr**: Clockwise (+) or counterclockwise (−) **Crr**: Consistency of refraction rotation (Rr) ratio (a number between 0 and 1)
	Angle of bifurcation (**Ab**)	**Ab**: Directly measured angle **Abf**: Relative frequency of Ab
	Rotation of bifurcation (**Rb**)	**Rb**: Clockwise (+) or counterclockwise (−) **Srb**: Sequences of refractions (R) and bifurcations (B) **Arbr**: Alternation of refraction (Rr) and bifurcation (Rb) rotations; node's charge

Appendix A further details this example regarding the quantification of the recursive irregularity of deviation, and how the data are presented and used. Reflecting on the relative virtues of statistical calculations is beyond the scope of this book; instead, we rely on collaboration with experts in the field. Observations and measurements are organized according to the four hierarchical levels presented in Chap. 6 (Figs. 6.23–6.29). Therefore, the parameters extracted from the observations, measurements, and statistical elaborations of the analyzed patterns (summarized and elaborated as shown in Tables A.1–A.4 in Appendix A) define a system (the algorithm, Fig. B.4 in Appendix B) in which behavior generates the "visceral quality of space" (taken just as an example). This is what the model can test in relation to reproducing the spatial features of the existing urban situation (as shown in Appendix B, Fig. B.5).

This example provides evidence of the computational nature of urban patterns expressed in the quantified behavior of recursive irregularities, or their essential qualities, and in their statistical calculation and definition of parameters. The measurement and calculation process also revealed that the qualitatively observed features of the urban patterns are coherent with what the quantitative logic and parametric relationships highlighted as the main features of the system.

7.3 Modeling That Reflects the Specific Urban Phenomena

Historically, the "model" has carried different meanings. During the Enlightenment, it indicated a perfect example to imitate, in contrast with the concept of "type," which for Durand in teaching architecture indicated a basic scheme containing a set of geometric, distributive, and constructive properties. Under this understanding, the type was an orienting model, to adapt rather than copy and to support decision-making for design (Saggio 2007, p. 78). This was the beginning of rationally generated typologies, through which the abstractions of the modern age were ordered. The objectivity of the principles of the modern period was expressed within the theoretical context of the "manual," which was supposed to provide, with scientific clarity, functional solutions for objectively defined needs. This functionalism was an inductive method situated in a framework of strong theoretical concepts and a world based on certainties. Conversely, we are now increasingly living in an uncertain and indeterminate world full of infinite potentialities. This condition requires us to start with questions and hypotheses that can be verified and to act deductively (Saggio 2007, pp. 79–80). Solutions may be affirmed only through a process. According to Saggio, the earlier ideas of the model are increasingly being replaced by a more scientific understanding in which models are characterized as dynamic interconnections of information. These interconnections simulate but also generate in an evolutionary manner a project in which information becomes the main tool, and the contained data are not rigid but easily modifiable. This pushes the designer toward a "simulation philosophy" (pp. 79, 82), intimately correlated with the abovementioned deductive method. In this concern, an important instrument is the prefiguration of the procedural approach that implies the formalization of interrelationships between parts of topological nature, expressed in diagrammatic form. These relationships start from "DNA as a generative and regulative code" and pervade and shape the architecture (e.g., parameters presented in Appendix A). The results depend on a series of variables that intervene to evolve the diagram in one or another formal direction (Saggio 2007, p. 90).

The steps outlined in the previous section help to observe, analyze, and define a system (interrelationships) that behaves in a specific way. The approach focuses on understanding the human properties that give rise to specific urban phenomena, with the aim of using this understanding in the design process. In relation to that, the challenge stays in designing a generative process using information extracted from this specific reality (i.e., generative and regulative DNA like in procedural approach); and process behavior self-generates a new emergent reality that remains indeterminate as it interacts and adapts to changes in its internal and external environment. This self-organization tends to self-regulate in a cyclical casual fashion, based on a combination of bottom-up and top-down individual and group behaviors generated within the hierarchical structure of the system, out of which the system reproduces itself.

If we want the model to be adaptable to change and to self-generate, it must be able to "understand" and develop using a similar cognitive process to that of living organisms. As Maturana and Varela (in Bitbol and Luisi 2004, p. 99) explained, the cognitive process is the basis of living beings' interactions and adaptations to self-produce. These authors coined the term "autopoiesis" to describe this phenomenon (Hallowell 2009, p. 152), which is the basic feature distinguishing life as autonomous systems that auto-generate and interact with their environment through "structural coupling" as a "history of recurrent interactions" (Maturana and Varela in Hallowell 2009, p. 149), which affect the "ontogeny" or "history of structural changes" in living beings (Maturana and Varela in Hallowell 2009, p. 149). Like life itself, the model can be not only "autopoietic" but also "ontogenic"; that is, like a living being, it can bear the history of structural changes and so also carry memory. "Digital preservation" can play this role in a model.

In the framework of architecture, the term "autopoiesis" was used by Schumacher (2011, p. 168) to refer to systems of communication within architecture that serve society as "communicative frames" for interaction. In fact, he based his "unified Theory of Architecture" on architectural autopoiesis, based on the assumption "that architecture is an integral system of communications … that reproduces itself" (2011, p. 172) in "communication structures like key distinctions, concepts, codes, values, methods and media" (2011, p. 4), which constantly combine among themselves. The successful communication of architecture requires a comprehensive theoretical systematization to position architecture's challenges in a contemporary context and, crucially, to organize the complexity of a networked society. Within this framework, Schumacher also introduced the need to upgrade design methodology in response to opportunities created by new technologies. He termed this "parametricism" or "the parametric paradigm and the formation of a new style" (Schumacher 2012, p. 617), and saw this as offering the potential for a unified (2012, p. 642) "epochal" event" for the twenty-first century, rather than a "mere fad or fashion" (2012, p. 622). However, while he drew a clear distinction between "epochal styles" and "transitional styles" (2012, p. 644), he remained within a quasi-"monumental" use of parametricism and the sublimation of technological monuments.

Used in a more profound way (computational instead of parametric), Di Raimo (2010, p. 115) suggested that "autopoiesis could be the main referent phenomenon that organizes the building's cognitive pattern." This would result in architecture with a "cognitive body network with autopoiesis tendency," which would "move to developing a living being phenomenon." In this regard, architects like Roche, who has been at the forefront of these themes since the mid-1990s, have explored and created a different path in architecture within the information paradigm using a computational approach to approximate living systems and trigger automatic operations linked with body activity algorithms. His methodology consists of the "expansion of meanings in manipulating information" (Di Raimo 2014, p. 13) in the design process, which is far from any "linguistic speculation" enabled by the capacity of computers, or the "parametrization" of form seen with some important players in the field of architecture, such as Peter Eisenman and Zaha Hadid, during the first phase of the information paradigm. Instead, Roche attributed to information not only the role of formal configuration, but also of an "important organizational aspect (2014, p. 13)." He observed that form itself is subject to other dimensions of interaction with the environment and is in a "state of transformation," touching on the core of the information paradigm (2014, p. 13). Roche and his group (2014, pp. 13, 93) have proposed an "architecture that is self-produced," based on the regenerative "production and destruction" of architectural components using robots. In this sense, an "autopoietic machine is a machine organized … as a network of processes (transformation and destruction) that produces components which: (I) through their … interactions … regenerate and realize the network of processes … that produce them; and (II) constitute it (the machine) as a concrete unity" (Maturana and Varela in Di Raimo 2014, p. 93). This would be an autopoietic architecture open to all possible negotiations, in which we would find "science coupled with aesthetics, ecology with politics, and simulation with narration" (Di Raimo 2014, pp. 74, 11).

To summarize what has been argued thus far, the natural city tends to be autopoietic: it adapts, changes, regenerates, or degenerates based on intense communication and information flow, transcending time and space. These automatic actions, based on the internal intelligence raised during the self-organized processes, are much more likely to happen in the natural city than in one that has been designed from the top down. This begs the question: can the observations, analyses, and information extracted from the natural city (when adopting a holistic approach) be reified, concretized, and materialized in a model like life itself?

By modeling the reality, we can see that people are unconsciously part of a statistical model that can be expressed through quantitative indicators representing the most characteristic essence of a pattern. However, unconsciousness, in this case, does not mean people are not aware of how they create their own site-specific conditions. Here, collective intelligence or awareness emerges out of complex interrelated processes; it is embodied in the category of modality according to which a social group acts and gives feedback on individual behaviors through a cyclical process already explained. If durable, this complex and shared behavior, representing group intelligence, becomes a tradition or social norm. Detecting this intelligence as part of the pattern generation process, and translating it into input for a model, is an important issue. The correlation categories of motive, behavior, and modality, as the underlying factors of the essential quality of space, facilitate this detection and improve our understanding of the generative process. Thus, the essential quality at the base of a specific pattern also expresses collective intelligence; that is, the cognitive base needed to load an autopoietic model.

Having identified a methodology to influence conventional bureaucratic planning, the concern becomes how to load the information extracted from reality into a model. This is the subject of Appendixes A and B.

7.4 Conclusions

This chapter outlines a methodology to observe and analyze the urban patterns by applying conceptual bases of quantum, fractal, and complexity theories; and to organize the resulting information into a scientific model which embodies the characteristics of the observed phenomena.

Considering the essential qualities of the urban pattern as the containers of the information for the generation of these qualities, I start by explaining three methods of "measuring" these qualities as complex systems: complexity as algorithmic information content, statistical complexity, and complexity as degree of hierarchy (Mitchell 2009, pp. 13, 94–95, 111). These methods correspond to the nature of the described urban phenomenon, such as the repetition of certain actions creating discernible overall patterns, the frequency of occurrences that stand at the basis of repetition, and the hierarchical nature of the urban phenomenon made up of subsystems similar to fractal self-similar patterns.

The presented methodology consists of three main steps: The first one is related to the conception of the reality as a larger whole (quantum) that contains an underlying order made of relationships and hierarchy (fractal), and is self-organized as a collective whole based on individual/group interactions (complex). The keyword for this step is holistic reality. The second step is related to the application of the above lenses to unpack the essential qualities emerged as interrelationships generated by overlaps of past and present events (quantum); as recursive irregularity of elements repeated in a certain order (Fractal); and as cyclically reproduced individual/group interactions (Complex). The keyword for this step is unpacking interrelationships through analytical categories (Table 7.1). In the third step, the reality conceptualized from a holistic and relational perspective is translated into information to generate a model for better observation of the reality, or for design purposes. Sequential steps for depicting the recursive irregularities characterizing patterns' essential qualities and to translate those qualities into information (that define an algorithmic system) are described as the finalization of observation and analytical process (Table 7.2). The keyword for this step is translating emergent quality into information.

The essential quality, emerging from a cyclical self-organized generative process, contains the collective intelligence, or the cognitive base to load a model. Generating a model, which considers human behavior in a specific reality, stands to have repercussions for conventional and bureaucratic planning. In relation to that, some important issues to be further investigated through the modeling process are: What are the ranges of forms, beyond the observed ones, relative to the specific essential qualities of space (i.e., visceral, involute, collapsed, and inflated)? What are the informative parameters and their effects in relation to the essential qualities of space? These processes of Testing through models (e.g., the Historical Organic model or Recording Over model) which interfere with the city's ordinary regulative planning and design processes, may help to find the best fit for the official planning process within the behavioral model of society; that is, it may support the identification of regulations, tactics, and actions that fit the specific patterning models of Tirana. In relation to that, Silva speaks about tactical urbanism as a set of actions that fit within the concept of self-organization and may contribute to cities' evolutionary process in parallel with improvement of planning rules (Silva 2016, pp. 2, 7, 8).

This path urges courageous experimentation and the testing of an open framework of opportunities, with the strategy of myriad organic coordination and deregulation, while rationalized, not completely excluded, as in conventional planning and design. This is an added-value alternative to the top-down imposed mono-block building schemes. Such experimentations can explore the evolution of form and observe performance by modifying the informative parameters. If the model is interactive, relationships generated from a multitude of information can be observed, across a range of simulations. It can help to understand how it might perform in a continuously changing environment by changing some of the factors but without canceling its main qualities. As mentioned in this chapter, if we want the model to be adaptable, ideally it must be able to develop using a similar cognitive process to that of living organisms.

References

Arida A (2002) Quantum city. Routledge

Batty M, Longley P (1994) Fractal cities—a geometry of form and function. Academic Press, London, San Diego, New York, Boston, Sydney, Tokyo, Toronto

Bitbol M, Luisi Pier L (2004) Autopoiesis with or without cognition—defining life at its edge. The Royal Society, pp 99–107

Di Raimo A (2010) Architecture as caregiver: human body—information—cognition, ACADIA life information

Di Raimo A (2014) François Roche heretical mechanism and living architectures of New-Territories.com. In: Saggio A (ed) The IT revolution in architecture series. EDIL Stampa, Rome

Dovey K (2012) Informal urbanism and complex adaptive assemblage. Int Dev Pla Rev

Hallowell R (2009) Humberto Maturana and Francisco Varela's contribution to media ecology—autopoiesis, the Santiago school of cognition, and enactive cognitive science. Proc Media Ecol Assoc 10:143–158

Holland JH (2014) Complexity—a very short introduction. Oxford University Press
Jones P (2019) The case for inclusion of international planning studios in contemporary urban planning pedagogy, sustainability
Kamalipour H, Peimani N (2019) Towards an informal turn in the built environment education—informality and urban design pedagogy, sustainability
Marshall S (2009) Cities, design and evolution. Routledge, London
Mitchell M (2009) Complexity—a guided tour. Oxford University Press
Portugali J (2000) Self-organization and the city. Springer, Berlin Heidelberg GmbH
Saggio A (2007) Introduzione alla rivoluzione informatica in architettura, Carocci, Roma
Schumacher P (2011) The autopoiesis of architecture—a new framework for architecture, vol 1. Wiley
Schumacher P (2012) The autopoiesis of architecture—a new agenda for architecture, vol II. Wiley
Silva P (2016) Tactical urbanism—towards an evolutionary cities' approach? Environ Plan B: Plan Des
Silva P (2018) Designing urban rules from emergent patterns: co-evolving paths of informal and formal urban systems—the case of Portugal. IOP Conf Ser: Earth Environ Sci 158 012001
Zohar D, Marshall I (1994) The quantum society—mind, physics, and a new social vision. Quill William Morrow, New York

Chapter 8
Conclusions

Abstract In this chapter, I recall the main arguments developed in the book, with the aim of highlighting how such a contribution can change the way we look at and think about cities and how the proposed methodology could produce an impact, not only in education but also in the way cities are observed and analyzed. Throughout the book, I argue about the need for a new way of looking, observing, analyzing, and transforming cities by applying the lenses of quantum, fractal, and complexity theories. In relation to this goal, I propose a methodology that may potentially contribute new knowledge in the way we look at cities and how we understand the process in which complex systems interact and exhibit emergence. Finally, using Tirana as the case study, I try to answer the following questions: How can this methodology be reflected in terms of new knowledge of the city and in the current debate on Tirana? and How can the proposed methodology open a debate beyond Tirana regarding the ways in which cities are typically analyzed and understood? The main conclusions of the book are summarized in relation to the proposed methodology and analytical categories to observe and analyze patterns as self-organized emergent realities; the role of the people in an open system; urban planning and design practice that supports co-evolution; and education that focuses more on relationships than on objects.

Keywords Emergent realities · Co-evolution · Urban planning · Design · Education

Throughout the book, I argue about the need for a new way of looking, observing, analyzing, and transforming cities by applying the lenses of quantum, fractal, and complexity theories. In relation to this goal, I propose a methodology that may potentially contribute new knowledge in the way we look at cities and how we understand the process in which complex systems interact and exhibit emergence. In this chapter, I recall the main arguments developed in the book, with the aim of highlighting how such a contribution can change the way we look at and think about cities and how the proposed methodology could produce an impact, not only in education but also in the way cities are observed and analyzed.

Finally, using Tirana as the case study, I try to answer the following questions: How can this methodology be reflected in terms of new knowledge of the city and in the current debate on Tirana? and How can the proposed methodology open a

S. Dhamo, *Understanding Emergent Urbanism*, The Urban Book Series,
https://doi.org/10.1007/978-3-030-82731-1_8

debate beyond Tirana regarding the ways in which cities are typically analyzed and understood? The main conclusions in relation to these issues are summarized below.

8.1 Analytical Categories to Observe and Analyze Emergent Qualities

The methodology proposed in Chap. 7 operates through analytical categories that help unravel—in a holistic logic (and not as a mechanical disaggregation)—the complexity of self-organized urban patterns. These categories try to comprise and explain the full complexity of the phenomena involving material and non-material factors embodied in human relationships as well as the essential qualities of urban patterns. Not without any reason, I started this book by defining the yet unexplored urban condition of the natural city, which—as opposed to the city planned from the top—is typically neglected and underestimated, even though it is the most complex arena of human interrelationships. In this regard, I highlight the fact that, as humans, we tend to exclude a phenomenon when we do not fully understand it. As part of this discussion, I drew on notions from the scientific world with applications to the urban planning and design field, all of which are centered around the concept of interrelationships as the underlying bases for understanding reality in a holistic way. In this sense, the urban patterns are seen under the three main lenses of quantum, fractal, and complexity, respectively, as the unbroken web of overlapping relationships in the space and time continuum, as relationships or recursive behaviors of elements assembled in hierarchies that generate the internal or invisible structure of form, and as a self-organized relationship within a collective whole that tends to self-regulate through interacting (signaling or information processing) in a cyclical process. The three ways complement one another in understanding interrelationships and obtaining a more holistic picture of emerging urban patterns.

Once we understand the difference between mechanical reality and the holistic one, with the help of quantum, fractal, and complexity theories, we further unpack this reality with the help of correlation factors (see Sects. 6.1.3 and 6.4) and other categories that define them. The correlation factors of Motive, Behavior, and Modality are the connecting thread between the three complementary ways of understanding relationships in human-based, self-organized macroscopic behaviors. These factors derive from the holistic conception of reality and explain the triggering and putting in motion of the entire system. They are further defined by some analytical categories, such as Event and Event-horizon (see Sects. 3.2.1 and 6.2), which specifically connect and correlate the non-local temporal reality with the local temporal one within the web of wholeness (historic, cultural, anthropological backgrounds, etc.); the associated wave effect of the event transcending space and time and emerging as a motive in local temporal reality; and the interfering patterns in local temporal reality, underlying the behavior at the base/bottom of the system and its transmission through upper levels. This network or assemblage of interrelationships generates a dynamic system structure (see Sects. 3.3.3, 6.1.2, and 6.3) that is further defined by fractal city categories, such as subassemblies made by elements assembled in hierarchies, and relations among them generated by the dynamics of the process and expressed in the recursive nature of repeating irregularities across scales. As seen in Sects. 6.1.3 and 6.4, the self-organization categories of the control parameter and order parameter are inserted to better define the interactions between four levels of hierarchy and the emergence of partial equilibriums resulting from bottom-up and top-down signal-processing effect and actions, as well as the confirming and disconfirming of behaviors, through a cyclical self-organization fed back from the top.

Both the correlation factors and the control parameters are related to how parts of the system interact. While the correlation factors connect all the above-mentioned categories with the origin of the event, indicating historical and cultural reasons for the dynamics and structure of the system as well as the passage from individual to social behavior, the control parameter and order parameter further describe the details of the cyclical process (signals, permissive membranes, strengthening/weakening of signals, thresholds when the behaver is enslaved, etc.) within the system. This self-generated internal world of relationships also serves as an indicator of how awareness and intelligence are raised in the system and embodied in the essential qualities of the pattern from which we begin the observation and analyses.

8.2 Looking at Patterns as Emergent Realities

Having a set of analytical categories that combine material and non-material aspects is not merely a matter of jargon; rather, it is an important conceptual shift that extends our observation field and allows for a more holistic understanding of reality. Thus, the proposed methodology and analytical categories help detect and learn how we can "see" this interactive and hidden universe. This is generated and raised in a holistic dimension with non-local temporal limits that generate a dynamic structure and set of patterns in local temporal reality. In this regard, the knowledge provided for new lenses on how to look at and understand the city (Chaps. 3 and 6), as well as the methodology presented in Chap. 7, are concrete tools that can be used to analyze urban patterns, considering a wider range of factors, or to unpack complexities related to a quantum and fractal approach.

Unlike those under a mechanistic conception, patterns are considered emergent realities or manifestations of horizons caught in the non-local temporal web of wholeness (event–event horizons). Under this logic, even apparently static and local temporal phenomena, such as urban patterns or specific areas of the city, gain a holistic dimension. In other words, their "territory" expands beyond the traditional limits strictly defined in time and space, expresses relationships within a self-similar hierarchic structure or demonstrates recursive irregularities through which a "new" relational order can be seen, and contains the self-generating cyclical process of relationships as a collective whole, or as the capacity to self-organize, evolve, and give rise to a higher-order complexity. This is vital in how we can approach a holistic understanding of the essential qualities of Tirana's patterning models.

8.3 Self-organization at the Base of the New Emergent Realities

In relation to that, the emergence of the natural city is an important issue in this book. For this purpose, I introduced the concepts of "organic," "informal," and "sprawl" as expressions of natural cities that are complementary to each other (Chap. 2). Considering self-generation and self-organization as the two main characteristics at the base of natural cities and that of complex systems, this book attempts to understand how self-organized systems in the case of urban patterns and the related essential qualities emerge and evolve. This is achieved by investigating how a social structure based on a motive is set in motion during self-organization from the bottom up; how exchanging information and coordinating at the smallest level can generate an essential quality; how a self-organized system behaves at different levels; how its irregularity repeats throughout the system, thus creating recursive sequences and legibility; how intelligence and awareness raise and pervade the system through "filters" and permissive "membranes"; and, finally, why some behaviors and signals permeate the entire system but not others.

Using the logic of relationships within the internal structure of the self-organized processes generating urban patterns, we can also identify the specific geometry of the generative processes, or the dynamics leading to the new emergent qualities, apart from attributing structure to the analyzed phenomena. From this perspective, it is apparent that the natural city reproduces itself based on the intelligence emerging from interactions within the system. In relation to that, I discussed the potential reification of the qualities representing the auto-regeneration and intelligence of self-organized systems (expressed in the essential qualities of space) using a model that best represented these qualities. Potentially, this intelligence can be considered the autopoietic charge for a model that approximates life.

Accordingly, the application of the proposed methodology facilitates the investigation of interactions within this kind of self-organized systems from a holistic perspective and in a more systematic and detailed way. It also helps examine how such interactions create structure and patterns and how the latter are connected to the essential quality or morphological aspects of the patterns. All of these issues are addressed in more detail in the analysis presented in Chap. 6 and in the methodology presented in Chap. 7.

8.4 The Role of the People in an Open System

Intimately related to the topic of natural cities and self-organization is the role of people in shaping urban patterns. Although the book does not deal directly with this topic, it appears in relation to public participation, organized from the top through community structures, and in a more "immaterial" form that emerged from the bottom of the system triggered by a motive generated in the web of wholeness. In both cases, their activities are reflected in the physical reality. People can be part of the design process, not only through direct participation (as seen in the 1960s and 1970s), or making use in planning and design of their intrinsic logic in the patterning process, but also most importantly through enabling/guiding open systems to perform; that is, by balancing their "interference wave" pattern to act directly as "agents" in a developing process that combines bottom-up and top-down self-organized processes. That is, balancing a nuanced range of governance factors and institutions—either formal or informal—that guarantee learning as well as mutual and gradual co-evolution.

The New Organic pattern analyzed in Tirana (see Chaps. 5 and 6) is a typical example of how a self-organized open system can reach a temporary satisfactory equilibrium that guarantees a higher quality of life than do many centrally designed areas, with marginal support (in cases where it is given) from central institutions that do not destroy the system of internal relationships. Interestingly, the second version of this pattern—the Inflated New Organic—is an example of how a temporary equilibrium can become more "regulated" and satisfactory in the presence of some preexisting order or structure (e.g., agricultural infrastructure). This suggests that, rather than excluding and separating areas and people arbitrarily, official planning should support interrelationships both formal and informal—based on the specific reality.

As discussed, the so-called "natural city" is generated and exists only in the framework of an open system that enables all the relationships and connections within the web of wholeness. It can only survive if the system is not closed. To recognize this, we need to educate ourselves on the concepts of the holistic dimension or the unbroken web.

8.5 Methodology Supporting Co-evolution—Impact on Urban Planning and Design Practice

The proposed methodology, along with the reconceptions presented so far in relation to the analytical categories, our understanding of the urban patterns as holistic and emergent entities, the role of people within the framework of open systems, and our enabling to observe and unpack interactions raised within the self-organized systems, might have an impact on urban planning and design practice in some directions.

The abovementioned reconceptions are expected to refine the way we look at the city by using interconnected theoretical lenses all converging in complexity theory, and rethink the formal–informal conception as a non-dichotomous one (Dovey 2012, pp. 371, 385) but rather integrated with each other and with the city. Thus, the way we look at the natural city, not merely as an informal settlement devoid of historical, social, and cultural meaning, but as expressions of human properties and the arena of a complex web of interrelationships, is the first step to introducing new methodologies in urban planning and design.

Rethinking the formal–informal link as a non-dichotomous conception and transferring it into planning and design practice might lead to methodologies that allow gradual coevolution. In this regard, we can give life to our cities not only by implementing top-down operations, often imposed in a conflictual way but by creating conditions for a natural coevolution process through combined planned and unplanned actions and the involvement of actors with a vested interest in the process. This gradual coevolution requires a longer time and the awareness of formal planning institutions to achieve more flexible urban planning, where both planned and unplanned actions learn from each other and mutually transform through continuous adaptations.

In relation to that, Silva (2016, p. 8) argues about the role of tactical urbanism, based on actions, and its contribution to cities' evolutionary processes that may potentially converge with planning institutions. Tactical urbanism fits the concept of self-organization and is an example of a bottom-up process as a self-adaptive response to citizens' needs and an expression of city complexity (2016, p. 2). In this regard, the "living wall" presented in Figs. B.7 and B.8 in Appendix B is an attempt to accommodate unplanned actions in planning and city evolutionary transformation. As Marshall says, while the "evolutionist paradigm" allows purposive interventions for cities to arise, "the outcome is still somewhat organic in being complex and emergent" (Marshall 2009, p. 262). If we consider evolution, the role of designers and planners becomes a discrete part of the process (Silva 2016, p. 3). This is especially true for complex parts of the city, where numerous people are involved, and the terms are typically longer. The interaction with the unplanned, the tendency to integrate, the coevolving paths of informal and formal urban systems, and the co-dependent adaptions and intertwined solutions all require planning flexibility, governance-wise, to learn how to improve the planning rules (Silva 2018, pp. 1, 4, 9, 10). More importantly, this must be a continuous process.

From an urban design and planning point of view, the beginning of such a process should be preceded by information and details on community ties, human interactions in the processes, levels, and actors involved, essential qualities of space, and related qualitative and quantitative indicators. The methodology proposed in Chap. 7, which represents a holistic conception, becomes a crucial aspect of providing a wider range of data to be used in the planning, designing, or a modeling process. Among other things, this includes testing through a large variety of simulations that combine the spontaneity of small-scale individual actions (residential cells) with larger-scale top-down interventions (infrastructures or buildings), which might help generate a better understanding of a specific situation and find the best fit with reality and the formal/regulatory planning (Fig. B.12). In this regard, the design of generative processes based on parameters (information) that reflect the essential qualities of a specific reality, is complementary with the coevolutionary planning approach.

In addition, in regard to the influence on urban planning and design practice, the proposed methodology features a strong historical and anthropological background, touching upon important issues, such as how to transform organic or spontaneous phenomena in a city into quality urban design, how to give space to a system of apparently irrational rules within a rational system, and how to follow not only a top-down approach but also a truly bottom-up or combined process. Importantly, as shown in Chap. 7 and Appendixes A and B, the computational nature of urban phenomena is closely linked to the historical–anthropological background. They do not exclude each other but are, instead, mutually part of the unbroken web of wholeness.

8.6 Education—Greater Focus on Relationships than on Objects

I argued throughout the book about the importance of exposing architects and urban planners to complexity as a critical issue in understanding a city. As mentioned several times, the territory is no longer a physically limited entity; rather, it exists beyond its local time and space and does not develop in a linear way but evolves over time. This conception, as Silva (2020, p. 1) says, helps spatial planning and design rules focus more on relationships rather than on objects. Thus, according to this author (2020, p. 1), planning and design education must refresh these concepts and reflect this shift, which then helps deconstruct dichotomies (e.g., planned and unplanned).

In planning practice, this conception still wields little influence. Thus, as mentioned in the book, it is crucial to include in planning education the concepts of informal urbanism and informal settlement as part of global phenomena learned in a cross-cultural environment (Jones 2019, p. 14). Planning education must also include in urban design not only magnificent civic places but also ordinary environments of informality that, while using multi-scalar thinking, focus on micro-scale, self-organized incremental production of space (Kamalipour 2019, pp. 5, 8). Closely related to this is the need to expose planning and architecture students to the institutional panorama and the wide range of actors generating the governance vectors as the main driving force in achieving coevolution.

The methodology proposed in this book can fit a diversity of educational environments, because it is conceptualized from a complexity perspective that, at the same time, plays the role of the connecting thread within this diversity. From this perspective (complexity), planning problems are not so different (Silva 2020, p. 1). This helps to move the planning focus from formal–informal aspects of objects per se to relationships and balances generated, or the roles they play within each urban system (2020, pp. 12, 13). In this sense, the proposed methodology attempts to reintegrate the simplified approach to reality, which contributes more to its fragmentation than to its unification.

8.7 Tirana and Beyond

The approach presented in this book perfectly fits the current debate on urban developments in Tirana and represents an alternative to evictions in the large construction schemes sustained by developers. Combining the small-scale incremental housing schemes and the large-scale developments would be an appropriate strategy for Tirana, in the absence of affordable houses and lack of support from the government structures. The same is true under conditions where informal arrangements based on individual small-scale negotiations are more successful than the redevelopment culture based on land consolidation. Over the past 5 years, Tirana has undergone a new wave of top-down interventions, such as the extension of the main boulevard, the Tirana River Park, and new residential areas, which are projects that typically overlap with the so-called informal settlements areas. In fact, the new residential blocks represent nothing new in terms of urban planning and design. On the contrary, they are outdated top-down schemes, similar in logic to the blocks built during the communist regime.

Under these conditions, the methodology presented in this book may have some advantages. First, it offers a more realistic understanding of the complex phenomenon currently existing in the city, especially in the project areas mentioned above. Second, it introduces an evolutionary city approach by combining small-scale individual actions with larger elements of top-down planning (mainly supporting or guiding infrastructure, or building structures), leaving more room for gradual and mutual transformations of formal/informal through incremental developments. Third, using models, this proposed methodology tests a large variety of simulations generated through information extracted from observation and analyses of the study area, following the methodology presented in Chap. 7. Finally, this method learns how to balance an open system and facilitate its performance.

The examples presented in Appendix B and in Figs. B.7–B.11 illustrate how a supporting infrastructure (large scale top-down) might guide developments of growing residential cells, which, while spontaneous and coordinated at the smaller scale, are guided and organized at a larger scale. Although the example is schematic, it illustrates the idea of "guided" spontaneity in the framework of a tactical urbanism. Thus, the use of methodology, in this case, might help identify the best fit with the specific reality and formal planning.

Further experimental steps could try to put together the essential qualities of Tirana's patterns (Historical Organic, Recording Over, and New Organic), together with the rational one of the European and Italian matrix that coexists and intersects with the Ottoman one, as presented in the historical analysis in Chaps. 4 and 5. In this sense, the official institutions representing city planning need to open their minds to learn more from the reality of the city. Indeed, how to learn from the city itself is one of the most important roles of this methodology.

In this book, I use Tirana as a case study. However, as previously mentioned, a similar approach can also be followed for other cities or other specific conditions. The proposed methodology may also be particularly valid in a broader Balkan and European context in the framework of multiple cultures and ethnic societies as well as increasing trends of migration. In this regard, the methodology may be a valid tool in identifying the essential quality of space of the different cultures living in a city (often considered problematic) and in observing and analyzing not only within the segregated immigrant ghettos, but also in their maximum expression/potential represented by the original events in the countries of origin, carried by human imprints (horizons form non-local–temporal realities as part of the methodology). Furthermore, the proposed methodology can help us generate a large, formal variety corresponding to the essential quality of space that is not a top-down solution, but emerges within the web of interrelationships, as envisioned by methodology, and fits within the concept of self-organization (embodying characteristics/actions specific for each community). Finally, the methodology may also be a valid tool in exploring new directions of cross-cultural coexistence based on methodologies that allow coevolution and convergence with planning institutions.

In this regard, we can learn from some failures of the past. I refer to the cases of many housing complexes thought to be perfectly designed as the paradise of modern architecture. Ironically, they favored the ghettoization of poor and emigrant populations moving toward urban, social, and economic dysfunctions. Many emblematic housing complexes exemplify this phenomenon, including Pruitt-Igoe in St. Louis, US, which was built in 1954 and totally erased starting from 1972 because of the social collapse; the spectacular hexagonal-shaped residential blocks in Bijlmermeer, Amsterdam, Holland, built during the 1960s, became undesirable for the middle class, and attracted the influx of emigrants, before being partially demolished and replaced with smaller buildings starting from 1992; the *grands ensembles* in the *quartiers Nord* of Marseilles, France, built in the 1960s and subjected to rehabilitation where parts of the buildings were demolished; and the so-called *Muraille de Chine* in Saint-Étienne, France, built in 1965 and demolished in 2000, to name a few.

Among other problems due to the wider context in the respective countries, these neighborhoods, which were designed according to the universal principles of modern architecture, were destined to fail or experience problems, as the spaces they offered and the behavioral models of the people who inhabited them were incongruent with the prevailing cultural, social, and economic contexts. In this regard, having a methodology that considers—in a complex way—the specific realities, as well as provides information to generate models that consider the historic and anthropological features of a city, can be extremely useful. Specifically, such a model can be used in exploring and testing a wide variety of alternative solutions in the framework of healthier multicultural, multiethnic, and multireligious societies. Searching for new solutions should go beyond demolitions that consider the city primarily as a physical structure. This should be better balanced by the idea of the city as relationships staying at the base of new, emergent realities. These kinds of complex societies are becoming typical for many cities in the developed world and are expected to generate diverse challenges in the future.

Throughout the chapters of this book, I argue about the need for a more complex understanding of the reality in our cities from a theoretical and practical perspective. The investigation sought to answer the main question raised at the beginning of this book about new ways of looking at and understanding our cities from a more holistic perspective, which is related to our goal of understanding the correspondence between the web of human interrelationships and the emergent qualities of space. In this regard, the observation and analytical methodology presented in this book have broadened the observable base and brought into the sphere of consciousness excluded or not visible occurrences, or even realities that stay beyond territorialities understood in mechanical terms. In relation to that, it has been argued that the emergent qualities of urban patterns can only be detected if the theoretical lenses and analytical methodologies enable such discovery. Thus, throughout the book, I sought to provoke and enhance our senses to enable such detection, which is not a direct intuition, by using additional theoretical lenses. At the same time, I aimed to provide evidence that the information for city planning, design, and self-regulation is contained within the city itself. Only through this process can we understand the intelligent naturalness of human settlements and stop fighting it through laws and the technocratic conformity of bureaucratic planning.

References

Dovey K (2012) Informal urbanism and complex adaptive assemblage. Int Dev Plan Rev

Jones P (2019) The case for inclusion of international planning studios in contemporary urban planning pedagogy, sustainability

Kamalipour H (2019) Peimani N (2019) Towards an informal turn in the built environment education—informality and urban design pedagogy. Sustainability 11:4163

Marshall S (2009) Cities, design and evolution. Routledge, London

Silva P (2016) Tactical urbanism—towards an evolutionary cities' approach? Environ Plan B: Plan Des

Silva P (2018) Designing urban rules from emergent patterns: co-evolving paths of informal and formal urban systems—the case of Portugal. IOP Conf Ser: Earth Environ Sci 158 012001

Silva P (2020) Not so much about informality—emergent challenges for urban planning and design education. Sustainability 2020(12):8450

Appendix A
The Measurement Process

This appendix provides more details on the measurement process and on how the obtained information can be summarized and organized in a scientific model. The sample area 1.1 (Figs. 5.1 and 7.2), from the Historical Organic pattern, is used as an example to illustrate this process. In this regard, an attempt is made to quantify the recursive irregularity of deviation (Table 7.2). The process starts with the identification of the recursive irregularity (deviation) characterizing patterns' essential quality (Visceral), and elements defining the recursive irregularity (e.g., the nodes, lines, and angles); then we define factors indicating the behavior of the elements (e.g., number of entities, length, rotation, etc.) and the kind of measurements for each of the factors. This process includes absolute and relative values, ranges of distribution, strings of information, and the index numbers indicative of the behavior of a line.

Table A.4, which is a further detailing of Table 7.2, contains the statistical calculations that combine all relevant data from the previous steps and the key parameters that synthesize the essential qualities of the pattern (in this case the recursive irregularity of deviation). This information serves to define the algorithm as shown in Appendix B.

As mentioned before, the parameters and strings of information resulting from the measurements, and statistical calculations of the analyzed patterns define a system (the algorithm) in which behavior generates the essential quality of space (e.g., visceral, involute, collapsed, and inflated). In this sense, Appendix A is an extension of step 3 of the methodology presented in Chap. 7.

A.1 General Information

Ranges serve to distribute across relative frequencies the values registered in the measurement process. For example, the ranges used for the distribution of values in the case of segmental length (Ls) are as follows: 1–20 m, 20–40 m, 40–60 m,

S. Dhamo, *Understanding Emergent Urbanism*, The Urban Book Series,
https://doi.org/10.1007/978-3-030-82731-1

60–80 m, 80–100 m, 100–150 m, and 150–200 m. There is no range 1–10 m because no values were registered in this range. Similarly, the ranges used for the angles of refraction and bifurcation (Ar and Ab) are as follows: 1–10°, 10–30°, 30–60°, 60–90°, 90–120°, 120–150°, and 150–180°.

Relative frequencies provide important information of value for selecting which parameters should be used in the modeling process. After registering the absolute values of segmental length, or any other element, the relative frequencies can be calculated according to the respective ranges and their distribution analyzed according to the hierarchical level. From this, we can draw some conclusions regarding which parameters characterize the organizing principles and dynamics (behavior) of the elements generating the pattern and its specific quality of space. We have to remember that these parameters are not abstract numbers; they represent structure and dynamics set in motion by the self-organized complex behavior during the settlement process. Recalling "statistical complexity" (Mitchell 2009, p. 102), a model can be produced "based on observations of the messages" of the system.

Strings of information are used to represent the distribution of values according to ranges and levels of hierarchy. Strings of information are also used for some other data that express sequences of events, such as the sequence of refraction and bifurcation (Srb), which gives an idea about the complexity of the node, and the alternation of the refraction and bifurcation rotations (Arbr), marked by a clockwise or counterclockwise direction (+ or −). Once we have these strings, the set of regularities can be depicted and analyzed in relation to their combination with randomness. The strings of information are sequences, similar to the explanation given in the "complexity as algorithmic information content" method (Mitchell 2009, pp. 98–99).

Some other data are represented by an index number resulting from the ratio between different measurements. For example, the nodes per length (Nl) index results from the ratio of the total length of the line to the total number of nodes. (Refer to Table 7.2 for further examples.)

Nodes are created by refraction (a deviation from the straight-line perspective) and bifurcation (a point at which a new line starts and extends independently; Fig. A.1). Nodes can be simple or complex. Nodes are simple when they contain only one event, such as refraction (R) or bifurcation (B). Nodes are complex when they contain a combination of more than one event, such as refraction and one or more bifurcations. Each node has an identification number, such as 1, 2, 3, or 4. In addition, the node is marked by the events that created the node (e.g., by R and/or nB, where n is the number of bifurcations), and the rotation direction is marked with a plus (clockwise) or minus (counterclockwise) for simple nodes, or with more than one sign for complex nodes. For each bifurcated branch, numbering starts from the number of bifurcated nodes (Fig. A.1).

Fig. A.1 Nodes and lines

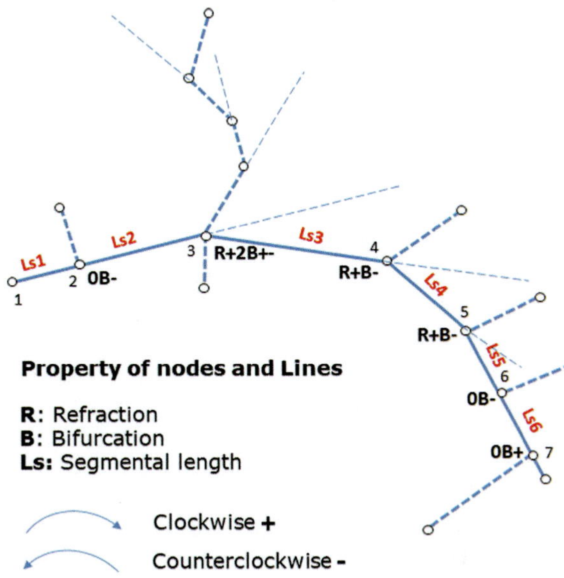

A.2 Elements, Factors, and the Measurement Process

The measurements in the following paragraphs follow the structure of the elements and factors presented in Table 7.2. These elements characterize the recursive irregularity of the deviation, typical of the visceral quality of the Historical Organic pattern.

Lines

> **Ls**: Straight-line segmental length; **Lsf**: Relative frequency of segmental lengths; **Lsi**: Segmental length index number; and **Lsfi**: Relative frequency of segmental length index numbers (Fig. A.1).

Segmental length (**Ls**) is measured by the distance between two consecutive nodes along the same line. Measurements are done separately for each line according to their hierarchical level, from the primary paths to the lower levels, as presented in Fig. 7.2. The observation and measurement of this phenomenon provide evidence for the relative frequencies of Ls and their behavior by level. The higher the level of hierarchy of the lines, the lower the relative frequency of the ranges with the shortest segmental length (1–20 m). For example, the relative frequency of the Ls range 1–20 m is 32% and 35% at the two higher levels of the hierarchy, while the frequency of the same range increases to 71 and 91% at the two lower levels of the hierarchy. Moreover, at each of the levels, this range occupies the main portion in relation to all other ranges. The frequency of the Ls range 40–60 m is close to stable at the two higher levels, at 21–26%, significantly decreasing at the two lower levels to 3 and 1%. The weight of the ranges starting from 60 to 100 m is almost insignificant. It is understandable that the range 1–20 m constitutes the main dimensional scale, considering our analysis is at the neighborhood scale and below. This means that 1–20 m is the most appropriate dimension (corresponding to the flexible module) for guaranteeing the required concealment for privacy. Further, if we combine the lower two Ls ranges to make a range of 1–40 m, we achieve an absolute majority, with frequencies of 64% for both of the higher levels, increasing significantly to 96% and 99% for the two lower levels.

The variation of Ls, from minimal to maximal absolute values within the same level of hierarchy, is as follows: for the highest level, $223 \geq Ls \geq 6.5$; for the second highest, $98 \geq Ls \geq 11.5$; for the second lowest, $82 \geq Ls \geq 1.5$; and for the lowest, $45 \geq Ls \geq 1.6$. Thus, the maximal value of segmental length reduces from the highest to lowest level of the hierarchy (excluding the length of 334 m in one of the primary lines as an outlier).

In Table A.4 is calculated the relative frequency (Lsfi) of the index numbers (Lsi) resulting from the ratio between the length of two consecutive segments ($Lsi = Lsn/Lsn + 1$). The application of the typical, or the most frequent indexes (Lsfi) as a "generator" on the "initiator" (the most frequent ranges of Ls), represents the logic of extension of the segments along the same line.

Nodes.

Nn: Number of nodes along the lines; **Nl**: Index number of nodes per length.

The number of nodes (**Nn**) is measured by the number of refractions and/or bifurcations located along the same line. The index number of nodes per length (**Nl**) is measured by the ratio between the total length of the line (**L**) and the number of nodes along the same line (Nn); for example, Nl = 6 m/node shows the average distance between the nodes.

Nn and Nl are measured for all levels of the hierarchy. These measurements provide evidence that the higher the level of hierarchy of line, the higher the number of nodes (Nn) per line. For example, at the highest hierarchical level, the Nn range is 12–39, while at the second-highest level it is 6–20. At the lower two levels, the range is 2–15 and 2–8. Further, lines with two to three nodes constitute 65% of the lines at the lowest level of the hierarchy and 46% of the lines at the next lowest level. This is understandable, because these lines serve a reduced number of houses in a private or almost private atmosphere. Lines with such a low number of nodes do not exist at the second-highest and highest levels of the hierarchy, which start from 6 and 12 nodes, respectively. This is also understandable, because these lines correspond to public or semipublic streets, which guarantee passage to higher levels of privacy.

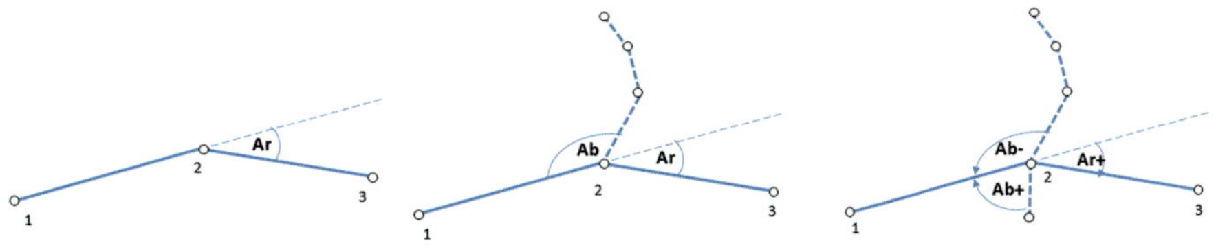

Node 2, case 1
Only refraction

Node 2, case 2
Refraction and bifurcation

Node 2, case 3
Refraction and bifurcation in two directions

Fig. A.2 Examples of possible refraction and bifurcation combinations

The Nl average is greater in the higher levels of the hierarchy. Starting from 29.2 m at the first level, it reduces to 18, 15, and 10 m moving down the levels. These figures align with the fact that the range of Ls from 1–20 m corresponds to the main neighborhood scale. At the highest level, the average distance of almost 30 m between nodes is sufficient room for a group of two to three houses (subsystems) with their gardens in first contact with this line. At the lower levels, the Nl reflects the measurement of a single house with its garden or even of a usable vain and the space for a connecting path. In this case, the potential to conceal and to create visceral space is higher. However, the visceral space also depends on other factors, such as the rotation angle and direction of rotation (discussed below).

Angles and Rotations

Ar: Angle of refraction; **Arf**: Relative frequency of Ar; **Ab**: Angle of bifurcation; **Abf**: Relative frequency of Ab; **Rr**: Rotation of refraction; **Rb**: Rotation of bifurcation; **Crr**: Consistency of refraction rotation (Rr) ratio; **Arbr**: Alternation of refraction (Rr) and bifurcation (Rb) rotations; **Srb**: Sequences of refractions (R) and bifurcations (B).

Ar, the angle of refraction, is measured by the angle of deviation from the straight line of the segment at the nodal point where this deviation starts (Fig. A.2).

Ab, the angle of bifurcation, is measured by the angle created at the nodal point between the bifurcated segment and the line from which the bifurcation starts.

As already explained, nodes can be simple (Fig. A.2, case 1) or complex (Fig. A.2, case 2), and still even more complex when bifurcation occurs in more than one direction (Fig. A.2, case 3). In this latter case, the angle of bifurcation is distinguished by the rotation direction. A plus sign indicates the rotation is clockwise, while a negative sign indicates a counterclockwise direction. In each of these cases, we can measure the refraction and bifurcation angles (Ar and Ab).

Based on our measurements, the variation of Ar, from minimal to maximal absolute values within the same level of hierarchy, is as follows: for the highest level, 94

$\geq$ Ar $\geq$ 1; for the second highest, 66 $\geq$ Ar $\geq$ 2; for the second lowest, 109 $\geq$ Ar $\geq$ 3; and for the lowest, 116 $\geq$ Ar $\geq$ 6. Thus both the maximal and minimal values of Ar are higher for the lower levels.

However, the values above represent the extreme margins. To obtain detailed information about how all other values are distributed through the ranges of 1–10°, 10–30°, 30–60°, 60–90°, 90–120°, 120–150°, and 150–180°, the relative frequencies of Ar (**Arf**) are calculated. This is a similar process to what I described in the case of Ls. From the measurements of Ar and their groupings by range, we see that the higher the level of hierarchy, the higher the relative frequency of the smaller angles of refraction. For example, at the first hierarchical level, the 1–10° range has a frequency of 68%, making this the main weight in relation to all other ranges. This then decreases to 53% at the second-highest level, before significantly decreasing at the lower levels, to reach 6 and 7%. The opposite is true for the two lower levels of the hierarchy, where the Ar range of 60–90° has a frequency of 50% and 45% of the total weight of ranges, respectively. These results make sense, as the higher levels of the hierarchy correspond to the more public streets, which are nearer to the straight-line perspective. By contrast, at the lower levels of the hierarchy, the higher values of Ar guarantee more privacy (at nearer to 90°, there is almost complete deviation).

In a similar way, we can study the variation of the angles of bifurcation (Ab). From minimal to maximal absolute values within the same level of hierarchy, the variations are as follows: for the highest level, 173 $\geq$ Ab $\geq$ 44; for the second highest, 155 $\geq$ Ab $\geq$ 74; for the second lowest, 165 $\geq$ Ab $\geq$ 54; and for the lowest, 104 $\geq$ Ab $\geq$ 11. Thus, the maximal values of Ab are higher for the higher levels of the hierarchy, while there is no distinguishable pattern in the variation of the minimal values. However, from the measurements of the bifurcation angles (Ab), the highest relative frequency of angles of bifurcation for all hierarchy levels is within the range 60–90° and 90–120°. Starting from the highest level of the hierarchy and moving toward the lowest, the frequency weights for these ranges, respectively, are 41% and 48%, 57% and 34%, 48% and 44%, and 44% and 50%. These figures demonstrate that the role of bifurcation is to maximize the concealing process. The fact that the highest relative frequency of the angles of bifurcation falls within the range of 60–120° means that the deviation from the straight line is almost maximal.

Rr, the rotation of refraction, and **Rb**, the rotation of bifurcation, are observed at the nodal points located along the communication lines. A plus (clockwise) or minus (counterclockwise) sign is allocated depending on the direction of rotation (Fig. A.3) as additional information to the measurement of the angle. Rr and Rb are presented as strings of information in the alternation of refraction and bifurcation rotation directions (**Arbr**), as explained below.

Crr, the consistency of refraction rotation index, is measured by the ratio of clockwise to counterclockwise rotations along the same line of movement at the same level of hierarchy. It shows the behavior of the line in relation to the quality of concealing. The division starts from the smaller number. Bifurcations are not counted as a change of direction because bifurcation starts a new line of movement at a lower hierarchical level. Crr can be equal to 0, equal to 1, or fall between 0 and 1, but it can never be bigger than 1. Examples are shown in Fig. A.3 such that:

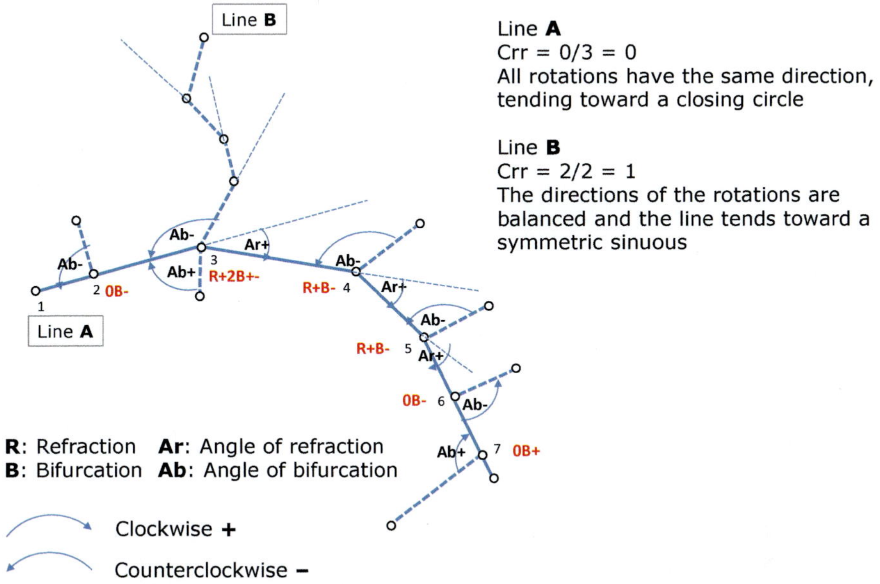

Fig. A.3 Rotation and the related parameters: rotation of refraction (Rr); rotation of bifurcation (Rb); consistency of refraction rotation (Crr); the signs of the black text show the alternation of refraction and bifurcation rotation directions (Arbr); the red text shows the sequence of refraction and bifurcation (Srb)

1. For line A: Crr — 0. If Crr — 0 or close to 0, all or most of the rotations have the same direction, tending toward a closing circle. The potential of this line to create concealing conditions is higher.
2. For Line B: Crr = 1. If Crr = 1 or close to 1, the directions of rotation are balanced and the line tends to be symmetric sinuous. The potential for privacy is also balanced.

The closer to 0 the Crr is, the more asymmetric sinuous the line tends to be. Thus, potentials to conceal and to create visceral space tend to be more similar to Line A. The closer to 1 the Crr is, the more symmetric sinuous the line tends to be, and the potentials to conceal and to create visceral space tend to be more similar to Line B. However, the potential to create visceral space also depends on the combination with other parameters presented in this section and with bifurcations.

Depending on the value of Crr, the behavior of the line in relation to the quality of concealing can tend toward a closing circle, toward being symmetric sinuous, or toward being asymmetric sinuous.

Arbr is the alternation of refraction and bifurcation rotation directions along the same line. After registering rotation direction information in strings for each line (e.g., for the line shown in Fig. A.3, the string would be −, ++−, +−, + −, −+, +), the regularities of the alternations along the lines can be depicted and written in the form of string sequences for each line or group of lines with the same

Table A.1 Example of two- and three-node strings. Type of line is shown on the horizontal, while specific lines are shown on the vertical

2 Nodes														
	Lines													
Type of line 1												**58%**	11 -	Type 1
	-	-	-	-	-	-	-	-	-	-	-	**32%**	6 +	Type 2
Type of line 2												**11%**	2 +-	Type 3
	+	+	+	+	+	+								
Type of line 3														
	+-	+-										**100%**	19 Tot	
3 Nodes														
	Lines													
Type of line 1	+	+	+	+	+	+	+							
	-	-	-	-	-	-	-							
Type of line 2	-	-	-	-	-									
	-	-	-	-	-									
												35%	9 - +	Type 3
Type of line 3	-	-	-	-	-	-	-	-	-			**27%**	7 + -	Type 1
	+	+	+	+	+	+	+	+	+			**19%**	5 - -	Type 2
												19%	5 + +	Type 4
Type of line 4	+	+	+	+	+									
	+	+	+	+	+									
												100%	26 Tot	

number of nodes (ranges), according to the level of hierarchy. Recalling the method of "complexity as algorithmic information content" (Mitchell 2009, pp. 98–99), any line can be composed of a combination of regularities and randomness, represented in the strings of information.

In the analyzed example, strings are written for each line, grouped by the number of nodes, and depicting repetitions as potential regularities with a certain frequency. From this, we see that up to the four-node lines, it is possible to depict regularities for the entire string; however, beyond this number of nodes, we can depict only sequences of regularities and how they combine within the string. Thus, how we depict regularities (for entire strings or sequences) is subject to the number of nodes.

For example, for the two-node lines (the shortest lines) in the analyzed sample, 58% of nodes have a counterclockwise rotation, 32% rotate clockwise, and 10% rotate in both directions at the same node. For the three-node lines, 35% contain the same combination "−+", 27% have the combination " +−", 19% contain "−−", and another 19% include "++" (Table A.1).

Starting from the five-node lines, the regularities can be seen in the relative frequencies of the most repeated sequences within the strings with the same number of nodes or within the defined node ranges. Depending on the repetition and their combination, different typologies of lines are apparent, such as lines containing combinations of longer sequences[1] and lines containing combinations of shorter sequences. This is typical for long string lines. As such, after depicting the regular sequences, the irregularities should be observed to determine how the regularities and irregularities combine in the string and whether any patterns can be discerned (Tables A.2 and A.3).

For example, in the observed 12-node lines, we can speak only about the relative frequency of sequences and their combination (Table A.2). In this case, 36% of

[1] Longer sequences are considered to contain more than two signs; shorter sequences are composed by only two signs.

Table A.2 Example of strings with 12 nodes showing the sequences of regularities and their combination with irregularities. The information strings for the specific lines are shown vertically. Sequences of regularities are highlighted in gray

12 Nodes						
sequence: - + - + +-		option 1		**sequence: -+**		option 2
-	-			-	-	
+	+			+	+	
-	-			-	-	
+-	+			+-	+	
+-	+-			+-	+-	
-	+			-	+	
+-	-			+-	-	
+	-			+	-	
+	-			+	-	
-	+			-	+	
+	-			+	-	
	-				-	

Table A.3 Examples of strings with 14–15 nodes showing the sequences of regularities and their combination with irregularities. The information strings for the specific lines are shown vertically. Sequences of regularities are highlighted in gray

14-15 Nodes											
Sequences: "++-+" "++-"			option 1	**sequences "+-+"**			option 2	**sequence "-+"**			optoin 3
+	+	-	+	+	+	-	+	+	+	-	+
+	+	-	-	+	+	-	-	+	+	-	-
-	-	+-	+	-	-	+-	+	-	-	+-	+
+	+	-	-	+	+	-	-	+	+	-	-
-	+	+	+	-	+	+	+	-	+	+	+
+	+	-	--	+	+	-	--	+	+	-	--
+-	+-	+0	+-0	+-	+-	+0	+-0	+-	+-	+0	+-0
-	+	-	+	-	+	-	+	-	+	-	+
+	+-	-	--	+	+-	-	--	+	+-	-	--
+-	-	-	-	+-	-	-	-	+-	-	-	-
-	+-	+	+	-	+-	+	+	-	+-	+	+
+	-	+	+-	+	-	+	+-	+	-	+	+-
-	-	-	-	-	-	-	-	-	-	-	-
+	-			+	-			+	-		

the strings comprise repetition of the (long) sequence "−+ −+ +−" (option 1), and 18% have at least one repetition of the sequence "−+" The remaining almost 45% are "irregular." On the right of Table A.2, a second option is shown in which regularities are depicted based on the repetition of the "−+" sequence.

Srb, the sequence of refraction and bifurcation, indicates the alternation of the refraction and/or bifurcation nodes along the same line. In a similar way to Arbr, once the strings of information for each line according to the number of nodes have been obtained, they can be analyzed to identify the regularity of the alternations along the lines and expressed by string sequences for each line or group of lines with the same number of nodes, according to the level of hierarchy. Similar observations and analyses to those performed for Arbr can be elaborated for Srb.

A.3 Conclusions

The measurement process, described in some detail in this chapter, reinforces our understanding of the phenomena observed thus far. The quantitative indicators arising from this process will inform our modeling of behaviors in reality as a first step in creating a scientific model. As we know, the entire patterning process is characterized by repetitive behaviors, some of which are more frequent than others; these frequencies are now quantified in relation to the influence they have on the essential quality of space, which is underpinned by human properties triggered by specific motives. In this regard, as argued in Chap. 6, the diagrams representing the repertoire of motives in Figs. 6.5, 6.6 and 6.7 (the mental patterns) help to give meaning to these statistics and explain the origin of these configurations. Here, we are dealing only with the quantification of the specific behaviors (Table A.4).

Following this logic, the dimensional scale revealed from the measurements of sample area 1.1 belonging to the Historical Organic pattern is consistent with the required concealment for privacy (the main motive). As mentioned in the text, 1–20 m (corresponding to the distance between two consecutive nodes) is the most frequent and appropriate dimension to guarantee the concern for privacy as the main anthropological factor. Similarly, the number of nodes along the lines corresponds to the hierarchy levels and the way houses group, that is, the lower the hierarchy level (the nearer to the visceral core), the shorter the distance between the nodes and the greater it is the potential to conceal. Also, the consistency of the refraction rotation shows that the more coherent the line is in the refraction rotation (tending towards a closing circle), the higher the quality of the concealment. This behavior of the line in relation to the quality of the concealment is supported by what was previously observed qualitatively in the Historical Organic areas. As shown in this appendix, there are other examples that show this consistency. Thus, as should be clear, the quantitative analysis of the measurements presented is consistent with the qualitative analyses and observations presented in the previous chapter. Finally, the measurement process reconfirmed the presence of a computational nature in the recursive irregularity of deviation characterizing the visceral quality of the Historical Organic pattern.

Table A.4 (a) Summary of information extracted from measuring the specific reality of sample 1.1 and calculated parameters. For illustrative purposes here is shown only the summary for the first level of hierarchy. A similar summary table has been formulated for each of the three other levels of hierarchy. This summary is based on measurements of lines according to the level of hierarchy as shown in tables (b), (c), (d), and (e) representing respectively a sample of the first, the second, the third, and the fourth levels of hierarchy. Similar tables have been formulated for each line. The descriptions presented in this appendix are based on the information summarized in these tables. For abbreviations refer to Table 7.2

a (summary of information and calculated parameters, first level of hierarchy)

Lines	Ls (m)	Sort Ls	N. Event.	Lsf (%)	Ranges	Nn	Nn sort	N. Even	Nnf (%)	Ranges	Nl	Nl Sort	N. Even	Nlf (%)	Ranges	Ar	Sort Ar	N. Even	Arf (%)	Ranges	Ab	Sort Ab	N. Even	Abf (%)	Ranges
A-C	25	6.5				13					33.05					5	1				113	44			
	28	6.5					12					20.19				2	1				107	60	2	2%	30-60
	40.6	7					13	2	50%	10-15		21.81	2	50%	20-25	3	1				111	68			
	42.8	7					34					33.05	1	25%	30-35	5	1				98	70			
	44	9					39	2	50%	30-40		41.75	1	25%	40-45	3	1				99	71			
	52	9.7						4	100%			116.8	4	100%		2	2				124	71			
	54.6	10									Nl Ave	29.2				3	2				111	74			
	56.6	12															2				99	74			
	86	13														8	2				128	74			
C-C	167	14				12					41.75					1	2				85	75			
		14															3				115	77			
		14															3				80	78			
		14															3					79			
		15															3				94	80			
		15															3				105	80			
		16															4				91	83			
		17															4				100	84			
		18.7	18	35%	1-20												5				104	84			
		20															5				96	85			
		20															5				97	85			
		22															5				74	86			
		22															5				102	86			
		24															7				105	86			
B-B	65	25				39					20.19					46	7				75	87			
	7	27.5														22	7				90	87			
	9	27.5														31	8				93	87			
	9.7	28														8	8				96	88			
	12	28														16	8				108	88			
	13	32														9	8				78	88			
	14	35.6														4	8				91	88			
	14	38.3														7	9				97	89			
	15	40														7	9	32	66%	1-10	92	89			
	16	40	15	29%	20-40											3	11				94	89			
	18.7	40.6														84	16				68	90			
	20	42.8														94	17				122	90			
	22	44														17	17				102	90			
	24	46														5	17				86	90	35	41%	60-90
	27.5	52														27	20				88	91			
	28	53.4														8	22				97	91			
	35.6	54.6														1	25				90	91			
	40	55														7	27				94	92			
	40	56.5														1	27	10	21%	10-30	92	92			
	46	56.6														1	31				148	92			
	66	57.5	11	21%	40-60											25	32				60	93			
	80	66														8	35				130	93			
	223.3	80															46	3	6%	30-60	97	94			
		80	3	6%	60-80												84	1	2%	60-90	74	94			
		86	1	2%	80-100												94	1	2%	90-120	96	94			
		102																47	100%		120	94			
		114	2	4%	100-120																86	95			
		167	1	2%	150-200																71	96			
		223.3	1	2%	200-250																96	96			
			52	100%																	138	96			
																					89	96			
A-A	65					34					21.81					5					109	97			
	7															20					90	97			
	10															3					95	97			
	14															2					90	97			
	14															4					84	97			
	15															17					98	97			
	17															11					71	98			
	20															35					97	98			
	22															17					86	99			
	27.5															5					97	99			
	32															27					87	100			
	38.3															8					88	102			
	53.4															1					87	102			
	55															32					88	104			
	56.5																				87	105			
	57.5																				80	105			
	80															9					77	107			
	102															2					170	108			
	114															2					74	109			
																					94	111			
																					92	111			
																					173	113			
																					79	115			
																					84	120	41	48%	90-120
																					70	122			
																					85	124			
																					89	128			
																					83	130			
																					91	138			
																					88	148	6	7%	120-15
																					44	170			
																					89	173	2	2%	150-18
																					93				
																							86	100%	

(continued)

Table A.4 (continued)

b (measurements of line A-C: first level of hierarchy)

LINES	NODES Nodes along the line	SEGMENTS Measured segments	Ls (m) Segmental length (m)	Sort Soort Ls	N. event. Number of events	Lsf (%) Relative frequency of segmental length	Ranges	Ls (m)	Lsi inde Segmental length index number (ratio)	Sort Sort index number LS	N. even Number of events	Lsfi Relative frequency of index numbers	Nn Number of nodes along the line	Nl Nodes per length	Nodes Nodes along the line	Ar Angle of refraction	Rr Rotation of refraction	Sort Sort Ar	N.event number of events	Arf (%) Relative frequency of Ar	Ab Angle of bifurcation	Rb Rotation of bifurcation	Ab2	Rb2	Sort Sort Ab	N. even number of events	Abf (%) Relative frequency of Ab	Crr Consistency of rotation ratio (refraction)	Rrr Resultant of refraction rotation	Srb Sequence refraction & bifurcation	Arr Alternation of refraction and bifurcations rotatoins;
Lines	Nodes	segments	Ls (m)	Sort	N. event	Lsf (%)	Ranges	Ls (m)	Lsi inde	Sort	N. even	Lsfi	Nn	Nl	Nodes	Ar	Rr	Sort	N.event	Arf (%)	Ab	Rb	Ab2	Rb2	Sort	N. even	Abf (%)	Crr	Rrr	Srb	Arr
A-C	A	A-0	40.6	25											A																
	0	0-i	54.6	28	2	22%	20-40		0.7	0.5	1	13%	0.1-0.5		0	5	-	2											-5	OR	-
	i	i-iii	52	40.6					1.1	0.6					i	2	-	2			113	-			98				-2	BR	--
	ii			42.8						0.7					ii			3			107	-			99					BO	-
	iii	iii-iv	28	44					1.9	0.7	3	38%	0.5-1.0		iii	3	-	3			111	-			99				-3	BR	--
	iv	iv-vi	42.8	52					0.7	1.1					iv	5	-	3							107				-5	OR	-
	v			54.6						1.5	2	25%	1-1.5		v			5			98	-			111					BO	-
	vi	vi-vii	25	56.6	6	67%	40-60		1.7	1.7					vi	3	-	5		1-10	99	-			111				-3	BR	--
	vii	vii-viii	44	86	1	11%	80-100		0.6	1.9	2	25%	1.5-2		vii	2	-	8	8	100%	124	-			113	7	78%	90-120	-2	BR	--
	viii	viii-x	86						0.5						viii	3	-								124				-3	OR	-
	ix														ix						111	-			128	2	22%	120-150		BO	-
	x	x-xii	56.6						1.5						x	8	-												-8	OR	-
	xi														xi						99	-								BO	-
	xii														xii						128	-								BO	-
TOTAL			429.6		9	100%					8	100%	13				0+:8-		8	100%						9	100%				
														33.05														0	-31		

c (measurements of line b-b: second level of hierarchy)

Secondary	Nodes	segments	Ls (m)	sort Ls	Fe	Lsf (%)		Ls index	r sort	Lsfi	Nn	Nl	Nodes	Ar	Rr	Sort Ar		Arf (%)	Ab	Rb	Ab2	Rb2	Sort Ab		Abf (%)	Crr	Rrr	Srb	Arr	W	W	Wv
b-b																																
	30	30-27	62	14									30															BO				
	29			20	2	40%	den 20	0.23	0.2				29						74	+			74					BO	+			
	28			50					1.1				28						81	-			81					BO	-			
	27	27-25	14	56.5	2	40%	40-60		1.4				27	7	-	2							83				-7	OR	-			
	26			62	1	20%	60-80		2.5				26			5			91	+			85					BO	+			
	25	25-21	20					1.43					25	5	-	7		deri 10					86				-5	OR	-			
	24												24			7	4	100%	89	+			87					BO	+			
	23												23						86	-			88					BO	-			
	22												22						88	+			89					BO	+			
	21	21-15	50					2.50					21	7	-								89				-7	OR	-			
	20												20						83	+			89		60-90			BO	+			
	19												19						87	-			90	11	69%			BO	-			
	18												18						93	-			91					BO	-			
	17												17						91	+			91					BO	+			
	16												16						97	-			93					BO	-			
	15	15-13	56.5					1.13					15	2	-				89	+			97		90-120		-2	BR	+-			
	14												14						90	-	85	+	101	5	31%			2BO	-			
	13	Oka e fundit eshte llogaritur si bifurkim											13						89	-	101	+						OR	-			
TOTALS			202.5		5	100%					17				0+4-		4	100%						16	100%							
												11.91														0	-21					

d (measurements of line AC-iii: third level of hierarchy)

Lines	Nodes	segments	Ls (m)	Sort	N. event.	Lsf (%)	Ranges	Ls (m)	Lsi inde	Sort	N. ever	Lsfi	Nn	Nl	Nodes	Ar	Rr	Sort	N.event	Arf (%)	Ab	Rb	Ab2	Rb2	Sort	N. even	Abf (%)	Crr	Rrr	Srb	Arr
AC-iii	1	1-5	82					82							1																
	2			5.6	1	33%	1-20	24	3.4	3.4	1	50%	3.0-3.5		2						98	+			90	1	25%	60-90		BO	+
	3			24	1	33%	20-40	5.6	4.3	4.3	1	50%	4.0-4.5		3						90	-			91					BO	-
	4			82	1	33%	80-100								4						91	+			92					BO	+
	5	5-7	24												5	00									98	3	75%	90-120		00	00
	6														6					90-120	92	+								BO	+
	7	7-8	5.6												7	109	+	109	1	100%									109	OR	+
	8														8																
TOTAL			111.6		3	100%					2	100%	7				0-:1+		1	100%						4	100%				
														15.94														0	109		

e (measurements of line AC-iii-6: fourth level of hierarchy)

Lines	Nodes	segments	Ls (m)	Sort	N. event.	Lsf (%)	Ranges	Ls (m)	Lsi inde	Sort	N. even	Lsfi	Nn	Nl	Nodes	Ar	Rr	Sort	N.event	Arf (%)	Ab	Rb	Ab2	Rb2	Sort	N. even	Abf (%)	Crr	Rrr	Srb	Arr
AC-iii-6	6	6-7	14	4				14							6																
	7	7-8	4	4.5				4	3.5	0.5					7	87	-	87											-87	OR	-
	8	8-9	4.5	8.5				4.5	0.9	0.5	2	40%	0.1-0.5		8	90	+	89											90	OR	+
	9	9-10	8.5	9				8.5	0.5	0.9					9	94	-	90		60-90									-94	OR	-
	10	10-11	9	14				9	0.9	0.9	2	40%	0.5-1.0		10	89	+	90	4	80%									89	OR	+
	11	11-12	16.5	16.5	6	100%	1-20	16.5	0.5	3.5	1	20%	3.0-3.5		11	90	+	94	1	20%									90	OR	+
	12														12					90-120											
TOTAL			56.5		6	100%					5	100%	6				2-:3+		5	100%											
														9.42														0.67	88		

Appendix B
Patterns Emerging from a Set of Rules

This appendix aims to illustrate the production of a diagrammatic model working in a computational environment. This model is based on the observation, analyses, and organization of information extracted from area 1.1 (Figs. 5.1, 5.8a, 7.1) as part of the Historical Organic pattern shown in Chaps. 5, 6, and 7. As mentioned in Chap. 7, through the history of architecture, the use of a model in a design process aims to represent an organization of practices and methods that are useful for architects in exploring and managing the system of decisions related to the development of a project (Saggio 2015). Despite the different kinds of "models" used in design processes, during the experimentation presented in this appendix, I refer specifically to the diagrammatic model. This model prefigures a process and a series of possible relationships that define the development and evolution of the same model. In this way, the diagrammatic model passes through a series of iterations leading to its refinement after each consecutive step. With this in mind, the following examples are intended to be speculative investigations of how diagrammatic models—thanks to a computational approach—can manage complex issues in urban situations. Through this model, it may be possible to use a large quantity of information to produce numerous diagrams of spatial configurations. The large number of the model's possible outcomes is one of its characteristics, which allows the designer to explore more than one possible deterministic solution to complex urban issues.

Indeed, this kind of model aims to emphasize the highlighted emergent characteristics of urban complex systems. The diagrammatic model explored here has two distinctive features. The first one deals with the model itself as a scientific device capable of highlighting the main characters of the patterns coming from the previous investigation. Once built, the model could eventually emphasize dimensions that would otherwise be unobserved and not properly considered. The second feature concerns the type of scientific model, which is formalized as being within a computational framework and is particularly based on parametric variables and relationships. Working within a computational framework facilitates the creation of a highly formalized model. This model can efficiently work on geometrical issues related to the urban morphologies and can be further used for design purposes.

S. Dhamo, *Understanding Emergent Urbanism*, The Urban Book Series,
https://doi.org/10.1007/978-3-030-82731-1

The model has been created by using a well-known software environment, namely, the plugin for Rhinoceros©, Grasshopper, a visual programming language. This allows us to easily create algorithmic procedures and their prompt consequences in terms of geometry visualization. In this computational environment, the creation of geometries is governed by a sequence of consecutive instructions. The computer-generated model is then used to demonstrate that a certain pattern recognizable in a territory can be reproduced computationally and can even be used as the base to foster reflections and strategies in the creation of new urban configurations.

Regarding the modality of building a generative algorithm with Grasshopper, the majority of its components—Grasshopper's basic unity—has a set of inputs and a set of outputs connected by a series of wires. Building an algorithmic sequence means visually connecting one or more outputs with one or more inputs, thus allowing information to be sequentially processed both quantitatively and qualitatively. Despite this short explanation, the potential of the plugin becomes clear, especially when the geometry and the procedure are highly complex. Indeed, by bearing in mind the fact that every model is always a reduced and abstract representation of reality, I assume the fundamental hypothesis that an important part of the behaviors behind the urban transformations are recursive and that they tend to repeat themselves in a fractal way, as I tried to explain in Chap. 6. The needed recursiveness, together with the required loops for it to appear, has been created through Hoopsnake and Anemone, two additional plugins that can easily run information inserted in a sequential procedure and create a series of loops, through which the algorithm can repeat itself, as it should happen within a recursive procedure. Hoopsnake and Anemone then allow Grasshopper to perform recursive tasks; in this context, they can be used to design a model that can simulate fractal behaviors. Further, we will see how an initial algorithm can be consequentially developed and become the base of the creation of design proposals.

B.1 Random-Walk

I start by taking into consideration the Historical Organic pattern and some main characters I would like to emphasize in the model. One of them is self-repetition, which produces the pattern described in Chaps. 5 and 6.

The computational model we are considering has been built in two different phases. The first one aims to set up a logical procedure (recipe) for the recreation of some basic features observable in the Historical Organic pattern, and the second one aims to develop a more complex model for design purposes. The initial stage focuses on the setting up and fine-tuning of an algorithm that can recreate—through an iterative loop system—some fractal characteristics of the existing urban situation. In this regard, the first result is the establishment of a procedure aimed at reproducing one of the observed behaviors in the evolution of the Historical Organic pattern: the pseudo-random walk (Fig. B.1).

Only during the second phase, once the model is set up, can the values shown in Table A.4 be used as parameters of a more complex algorithm. This algorithm allows the advancement from a more abstract beginning to the schematic recreation of the existing urban configuration (Fig. B.2). The first phase focuses on the possible early stages of the creation of the urban pattern, which can be fairly approximated by a sort of random walk, capable of recreating the first path or trunk road (what could be read as a generative path for the urban settlement). In computer science, the random walk stands for the set of rules and procedures that creates a succession of segments, in which each consecutive piece starts where the precedent ends, to randomly develop in a new direction. In the first model, this is controlled by two parameters: the length of each successive segment and the angle of deviation from the previous one. From a set of numbers, coherent to the observable behavior in the territory, the algorithm randomly chooses the better values to calculate the generation of the random walk.

As shown in Fig. B.1, after several recursions executed by the algorithm in Hoopsnake, some consistent geometrical features start to become observable. For practical reasons, 10 recursions are considered sufficient in evaluating the behavior, which has become visible and approximate with the observed reality. In this case, the similarities in the same types of paths, common in Tirana, gradually emerge, and the algorithm is already capable of simulating spatial configurations.

This—not at all a granted result—could be a good base upon which to test the next steps of development as well as the model's ability to respond to a set of real parameters and data coming from the previous analysis and observations summarized in Table A.4.

The initial algorithm of the random walk in the next stage is further developed by incorporating the data taken by a series of surveys in the area object of the study. The intent is to have a more sophisticated model based on real values—one that can recreate and manipulate a larger range of behaviors and spatial configurations.

In this next computational step (Fig. B.2), three main zones are emphasized by three corresponding grey rectangles. Each of them represents a part of the algorithm focusing on some geometrical features: the angle of rotation among the segments (the upper one in Fig. B.2), the variation in the length of the segments (the zone in the middle), and the bifurcation points of the segments (the lower part). These three important blocks of the algorithm represent and control the abovementioned geometrical features observed in the Historical Organic sample area 1.1.

It is interesting to point out that a Gaussian graph is visible in each of the three blocks. These graphs plot a second-order function that can best approximate the calculated relative frequencies of the measurements taken from the observation. Moreover, the Gaussian distribution allows the achievement of more and less probable parameter distributions as observed in reality. The effectiveness of using the Gaussian graph is an important finding of the experimentation. It also represents an abstract representation of the repertoires of motives that are described and illustrated with their concrete shapes in Chap. 6, thus confirming the intuition that it is possible to transfer formal behaviors into mathematic expressions that describe them.

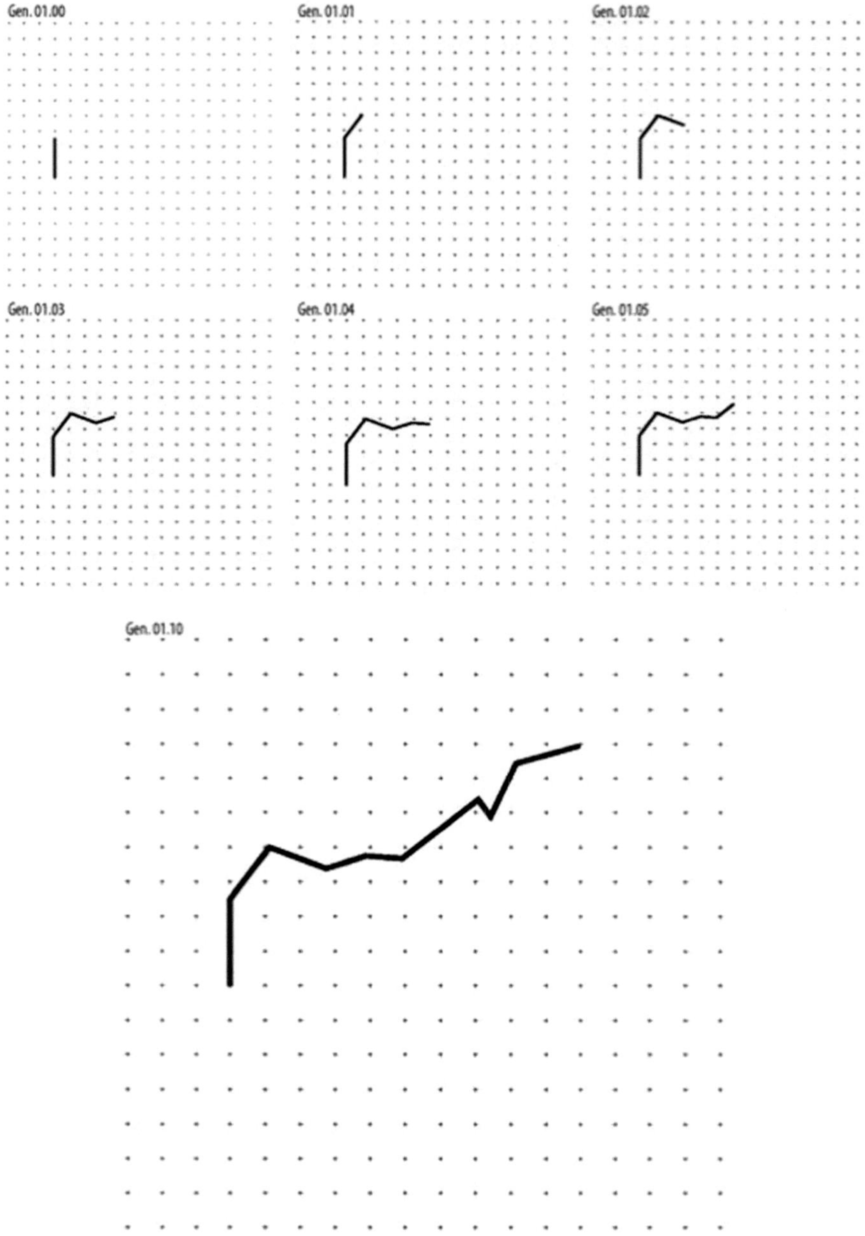

Fig. B.1 Pseudo-random walk

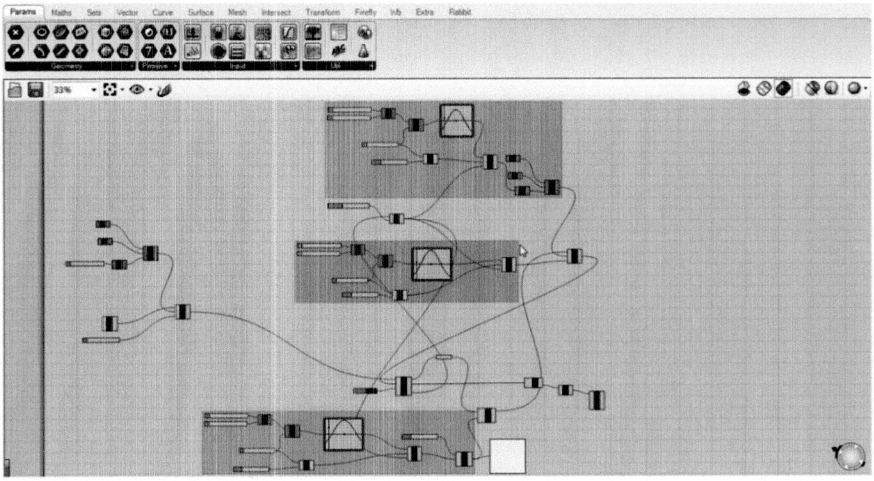

Fig. B.2 The complete, enriched Grasshopper script through empirical data

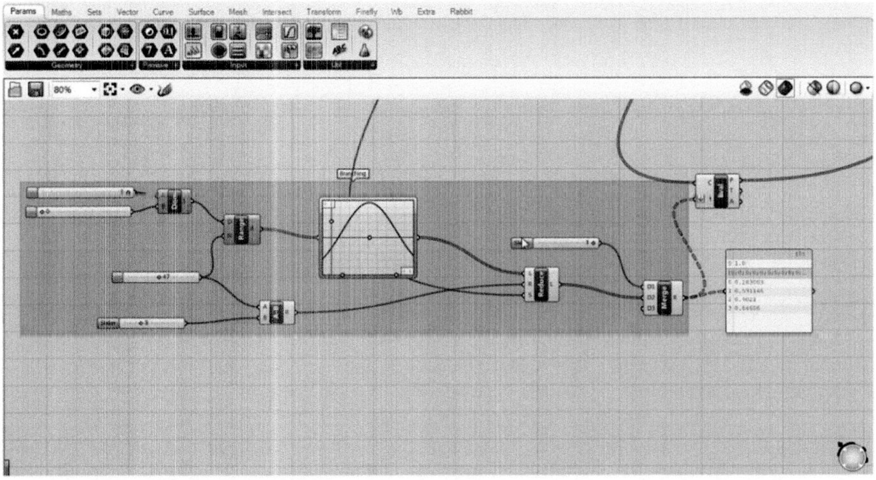

Fig. B.3 Zoomed-in image of the segment that controls the bifurcation system through a Gaussian graph

The lower part of the definition represents another crucial step in the procedure (Fig. B.3). This step defines the points at which the bifurcations start in every recursion. Within a set domain, which represents all possible points of bifurcation, the algorithm selects in each iteration only three values that identify three corresponding points on the segment from where the branches originated.

Indeed, when the above algorithm is applied, we can see the evolution of the random walk (Fig. B.4) informed by real parameters derived from the data shown

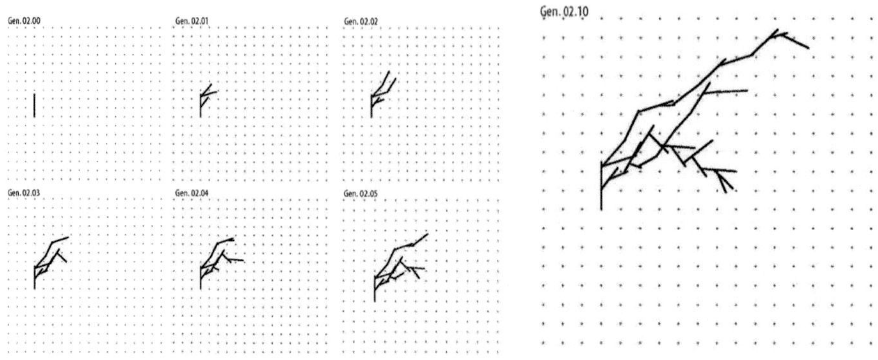

Fig. B.4 The second pseudo-random generation informed by a set of observed parameters and subsequent calculations

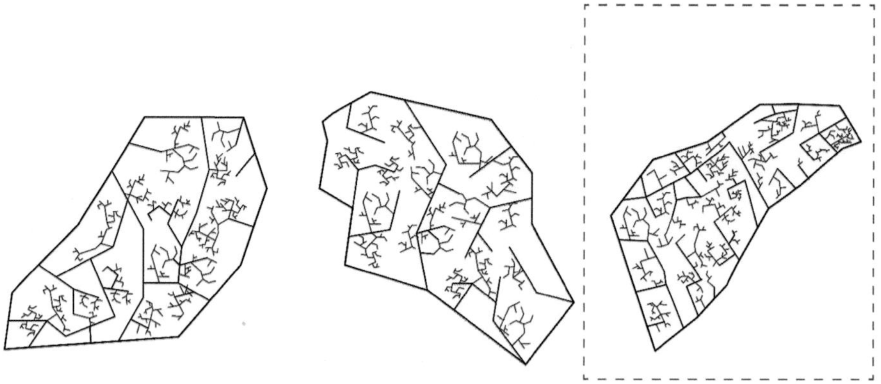

Fig. B.5 Three attempts to reproduce and manipulate recursive structures that are similar to the ones found in real urban situations; among these, the third diagram is the most promising in terms of similarity. Students: Aurora Çepele, Doriana Sherri, and Ensixhei Xhafaj in the framework of the EXP_LAB workshop

in Table A.4. To further verify the validity and effectiveness of the algorithm in reproducing the spatial features of the existing urban situation, we decided to simultaneously develop a larger number of branching geometries in order to create a more complex spatial organization. The algorithm uses as parameters the specific measurement taken from sample area 1.1. As shown in the results (Fig. B.5), the model created in the computational environment can reproduce and manipulate some recursive structures, which are similar to those found in real urban situations.

In fact, the iterative and generative algorithm comes close to recreating the complexity of the real conditions in the area being studied. If we compare the results (Fig. B.5) produced through the algorithm with the currently existing condition (Fig. 7.1), we can see a satisfactory similarity between the two. It is interesting to discuss that the information used in the model is, by definition, a reduced and abstracted version of reality. Nevertheless, the visible similarities confirm that this kind of model can recreate some existing conditions of the urban tissue and use the set of information to develop new spatial diagrams, which are useful for the next phases of design processes.

At this point, based on the results presented in Figs. B.4 and B.5, some further considerations can be made concerning the generative development of the Grasshopper algorithm. First, the necessary number of recursions to reach a certain degree of spatial complexity is lower than the simple random walk path that generally needs 10 recursions to produce a meaningful result. Notably, after five recursions, the branching becomes complex enough to recall some inner characteristics of the analyzed patterns.

Second, in the majority of the cases, the model is able to generate a satisfactory level of resemblance with the real conditions being examined (Fig. B.5). Therefore, we can assume the presence of a computational nature in some observed urban phenomena. In general, the intention of this experimentation is to claim that the formalization of the generative procedure through an algorithm is crucial for observing the emergence of morphologies that are visible in reality. The algorithm is also useful in understanding the evolution and possible variations of the whole system over time. The range and the distribution of the possible values for each parameter governing the algorithm allow for a closer approximation to the real conditions and further variations of the morphology. At this point, I present two examples rooted in the previously described concepts and models. These examples have generated specific design proposals starting from the analyses and measurements of sample 1.1.

B.2 EXP_LAB: Speculative Model to Foresee Emergent Strategies for Urban Development

Workshop held during the Tirana Architecture Week 2020 "'Science and the City" led by Sotir Dhamo, Ledian Bregasi, and Valerio Perna.

At this stage, it is important to explore the ability of the algorithm to allow variations during its development. This feature is the key to the future use of the model as a design tool that allows modifications, variations, and the emergence of novelty. Having a model that reproduces the historic and anthropological features of the city will be extremely useful for the future of Tirana. The gathered data could be transformed into meaningful patterns to shape a more informed development for the city. Meanwhile, as previously discussed in the historical analysis (Chap. 4), a European and Italian matrix coexists and intersects with the Ottoman and new organic ones.

The last section of this appendix aims to show how the diagrammatic model works in a computational environment and how the complexity of the information contained

Fig. B.6 Master plan with the reorganization of the area focused on neighborhood-based groups. The fractal characteristics of the urban pattern are recognizable in the spatial configuration. Students: Aurora Çepele, Doriana Sherri, and Ensixhei Xhafaj in the framework of the EXP_LAB workshop

in it can be used as a design tool. Groups of selected students were invited to join a number of intensive sessions wherein the research questions previously described were addressed, and the analysis phase was also transformed into a design strategy within the specific urban tissue of sample 1.1. Since the beginning, the approach with the students was non-solutionist. We did not aim to propose a deterministic model as a final result of the project; rather, we aimed to initiate a discussion in which the final diagram—and its consequent reification—was just one of the possible, and at the moment, most satisfying level of iteration reached.

The experimentation represented in Fig. B.6 was developed using this framework.

It is based on the observed tendencies (Chaps. 4–6) of the city's inhabitants to organize themselves in small neighborhood-based groups and to influence the spatial structure of the settlement through these groups. The algorithm is then used to define a strategy of land connection and occupation based on the fractal characteristics already observed in the area. The idea of urban farming through aquaponic technologies for fostering the internal production of goods and resources has become a driving force behind efforts to redevelop the area. The result is a proposal for the reorganization of the zone. In this proposal, the added urban layer presented in Fig. B.6 offers support to new technologies that guarantee the self-sustainability of the settlement while ensuring coherence with the historical structure of the neighborhood.

A second proposal focused on the concept of autopoiesis proposed by Humberto Maturana and Francisco Varela (Hallowell 2009, pp. 149, 152). As I have previously mentioned in Chap. 7, this concept refers to a system that is capable of independently reproducing and maintaining itself by reaching a balanced level of fitness within its inner components.

From a theoretical point of view, the autopoiesis idea permeated the work of this group. After understanding the current recursive feature of the urban tissue of the analyzed sample (Fig. B.7), they proposed the insertion of a living entity—represented in the proposal—as a branched tree trunk that could grow over time through the help of the inhabitants (Fig. B.8). This growth follows the same principles that gave birth to the existing urban structure. This is guaranteed by using the generative model based on the measurements. The tree trunk support structure could be intended as a technological device—one that is filled with a system of energy supplies that, as a proper living being, could inject substances and guarantee its survival. Not any single part of the tree trunk or "inhabited wall" would be wasted. Indeed, the thickness of the structure is used in several ways: at the lower level, it contains all the energetic infrastructure to act as blood vessels of the system; at the mid-level, we have the peoples' circulation; and at the top level, we have the trails that guarantee the free movement of a robotic arm (Figs. B.9 and B.10).

Later on, within these branches, a series of platforms were added. The latter would be a system wherein inhabitants could define the occupation/settlement strategy to realize their spaces and desires following a set of rules governing the density. These rules are rooted in the social consent of the Historical Organic pattern. In accordance with the decisions of inhabitants, the following "residential" system would be automatically realized by the robotic arm, which could freely move thanks to a system of trails deployed in the wall and, through the use of a material composed of a mixture of resin and *qerpiç*,[2] can realize the needed inhabiting structure by the citizens.

The strategy proposes an overall urban design within the tissue of the analyzed sample 1.1 (Figs. 5.1, 5.8a, 7.1). The strategy considers the emerging conditions and recursive peculiarities through an ever-growing structure that follows the fractal properties of the existing situations. The inhabited trunk/wall described in its inner

[2] Traditional construction material in the area of Tirana. The clay material from which qerpiç is made, before being turned into bricks or used in construction, is similar to a dough with a certain degree of liquidity or consistency.

Fig. B.7 Recursive fractal analysis of the existing condition of sample 1.1 and the initial diagram of the "living wall" (marked in orange) based on the measurement of the study area. Students: Flavjo Çyçi and Kevin Xhyra in the framework of the EXP_LAB workshop

features and in the way it is inserted within the neighborhood proposes new views regarding the building scale (Figs. B.9, B.10 and B.11). Finally, the proposal also covers the smaller system of residential cells. This system was created by the robotic arm following the behavioral decision of the agents (i.e., the inhabitants), who freely roam and populate the overall spaces based on the social consent explained in Chaps. 5 and 6. Thus, the final results show strong non-deterministic qualities. This result is not a fixed typology or option. Since the beginning, it has grown in accordance with the recursive features of the urban condition, and its further development is influenced by the inhabitants and their emergent behaviors.

Figure B.12 shows a combination of the top-down design approach with the bottom-up self-organization of the settlement. In order to do so, a supporting infrastructure of paths and public spaces creates the conditions for the densification of the area. The urban configuration is generated taking into consideration the parameters extrapolated from the analyses of the essential qualities of the specific pattern. The computational approach in this way complements the co-evolutionary theory in planning.

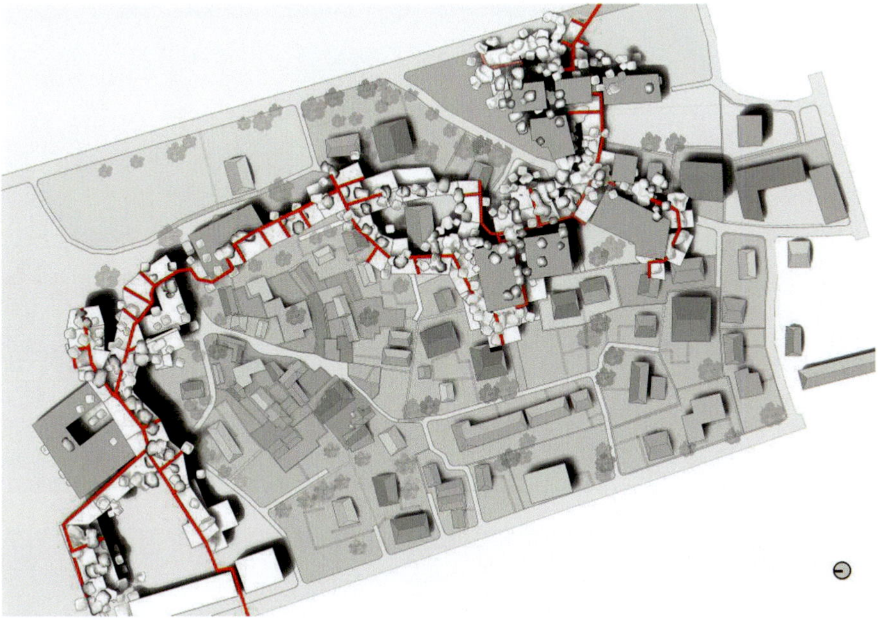

Fig. B.8 The final condition chosen for development after some computational design iterations. Students: Flavjo Çyçi and Kevin Xhyra in the framework of the EXP_LAB workshop

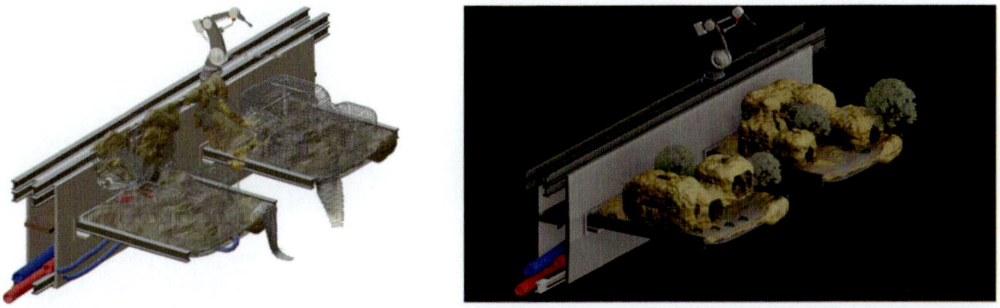

Fig. B.9 Details of a portion of the inhabited wall showing several features of the project: the trails that regulate the movements of the robotic arm, the inhabited thickness for the citizens' circulation, and the energy vessel. We can also see two different stages of the residential units on the platform: a "naked" skeleton and one of the possible configurations (materials: resin + *qerpiç*). Students: Flavjo Çyçi and Kevin Xhyra in the framework of the EXP_LAB workshop

B.3 Conclusions

The previously described examples can help in building up a number of considerations on the nature of the proposed interventions.

Firstly, we can notice that algorithms can recreate to some extends the spatial complexity of the urban condition. This implies that the possible sequence of events

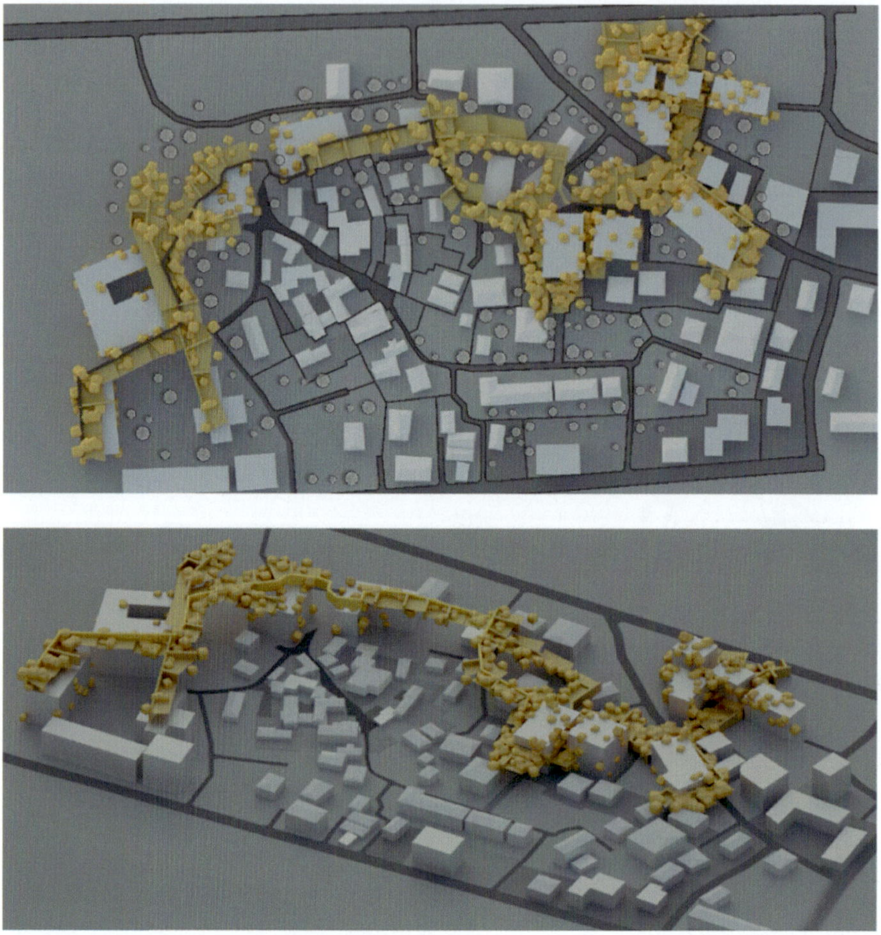

Fig. B.10 The master plan proposal; an axonometric view of the design proposal; and the support structure and spontaneously growing residential cells. Students: Flavjo Çyçi and Kevin Xhyra in the framework of the EXP_LAB workshop

that historically shaped the neighborhood and its essential qualities have been interpreted in a plausible way, within the assumption that any model is a reduction version of the reality. Thus, we can also assume that the sequential procedure of the algorithm embodies some of the essential qualities of the urban space. In this case, it can be affirmed that self-organization can solve problems by adapting its form to best fit the boundary conditions. Thanks to these kinds of algorithms, we can use the computational nature of some urban phenomena in order to formalize spatial behaviors into mathematical expressions.

Fig. B.11 Longitudinal sections, which show the relationships and interdependencies within the existing situation and the new "autopoietic" entity. Students: Flavjo Çyçi and Kevin Xhyra in the framework of the EXP_LAB workshop

Fig. B.12 Simulation that combines the spontaneity of small-scale residential cells with larger-scale top-down interventions (access infrastructure); Student: Flavjo Çyçi in the framework of the EXP_LAB workshop

The same algorithms can then be used as design tools. Their procedural nature allows the creation of spatial diagrams and the emergence of self-organized settlement models. These models can potentially fit within the real self-organized settlement generating principles and can be a supporting tool in simulating bottom-up small-scale adaptive tactics, in order to evaluate their contribution to city/neighborhood evolutionary processes, and how small-scale adaptations potentially meet larger planning objectives.

It can be noticed, for example, that the final results of the design process show strong non-deterministic qualities. These qualities are conferred by the way in which the inhabited wall (Figs. B.7, B.8, B.9 and B.10) grows and by the activities facilitated by its presence. The growth of the inhabited wall follows the recursive features of the existing urban condition. Its growth tracks the fractal recursive features observed in the neighborhood, allowing at the same time the emergence of smaller scale spatial organizations thanks to the enhancement of the interaction between the inhabitants. The emerged structures are negotiable through all of the scales (i.e., neighborhood, building, and cell) so the rendered images are just one of the many possible outcomes. The complex system generated by this process is capable of self-organizing in different spatial configurations, adapting itself to the ever-changing urban conditions. The ability of the system to maximize the fitness allows its survival and adaptation to the mutable factors such as the human presence, energy consumption and production, and material distribution.

The algorithm can be further useful in understanding and influencing the evolution and possible variations of the system in time. Thus, by applying variation to the parameters and sets of rules that govern the model, we can have an insight into how possible policies and projects can affect the transformation of our urban settlements within a coevolutionary logic. Thus, it can be an important tool for understanding the emergent urban phenomena and creating space for the emergence of novelty.

References

Hallowell R (2009) Humberto Maturana and Francisco Varela's contribution to media ecology—autopoiesis, the Santiago school of cognition, and enactive cognitive science. Proc Media Ecol Assoc 10:143–158

Mitchell M (2009) Complexity – A guided tour. Oxford University Press

Saggio A (2015) Da Alexander Klein a Ben van Berkel: dal modello oggettivo al modello diagrammatico per comprendere un capolavoro degli anni Duemila, in UNStudio diagramma struttura modello pelle ibridazione. Editors De Francesco G, Ghazi E, Santarelli E; Lulu.com: Raleigh, USA